Author:
Ruth M. Young, M.S. Ed.
Elementary Science
Consultant

Illustrators:
Sue Fullam
Keith Vasconcelles

Editors:
Evan D. Forbes, M.S. Ed.
Walter Kelly, M.A.

Senior Editor:
Sharon Coan, M.S. Ed.

Art Director:
Darlene Spivak

Product Manager:
Phil Garcia

Imaging:
Rick Chacon

Research:
Bobbie Johnson

Publishers:
Rachelle Cracchiolo, M.S. Ed.
Mary Dupuy Smith, M.S. Ed.

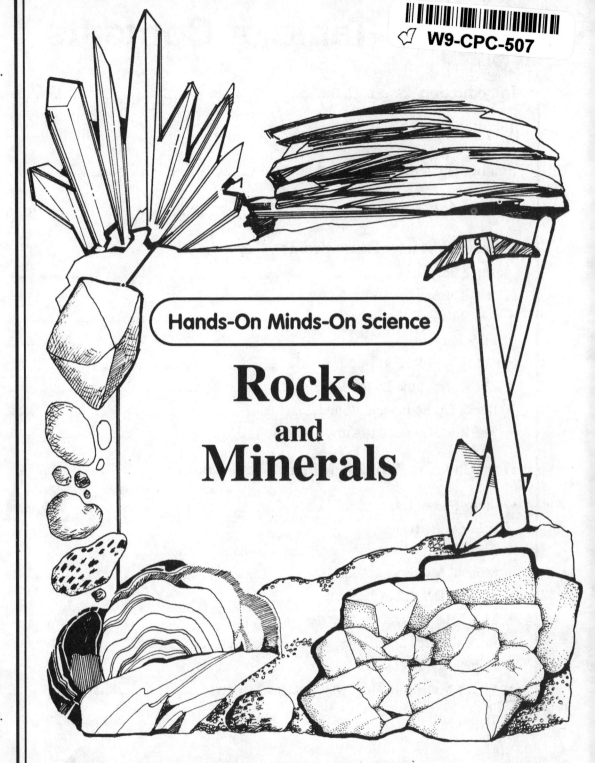

Hands-On Minds-On Science

Rocks
and
Minerals

Teacher
Created
Materials

Teacher Created Materials, Inc.
P.O. Box 1040
Huntington Beach, CA 92647
©*1994 Teacher Created Materials, Inc.*
Made in U.S.A.

ISBN-1-55734-636-4

Table of Contents

Table of Contents *(cont.)*

Introduction

What Is Science?

What is science to young children? Is it something that they know is a part of their world? Is it a text-book in the classroom? Is it a tadpole changing into a frog? Is it a sprouting seed, a rainy day, a boiling pot, a turning wheel, a pretty rock, or a moonlit sky? Is science fun and filled with wonder and meaning? What does science mean to children?

Science offers you and your eager children opportunities to explore the world around you and to make connections between the things you experience. The world becomes your classroom, and you, the teacher, a guide.

Science can, and should, fill children with wonder. It should cause them to be filled with questions and the desire to discover the answers to their questions. And, once they have discovered the answers, they should be actively seeking new questions to answer!

The books in this series give you and your children the opportunity to learn from the whole of your experience—the sights, sounds, smells, tastes, and touches, as well as what you read, write about, and do. This whole science approach allows you to experience and understand your world as you explore science concepts and skills together.

How Did Rocks and Minerals Originate?

Our planet was not always as we see it today. Most scientists think our solar system began about 4.6 billion years ago as a huge spiraling nebula (cloud of gas and dust-sized pieces of rock and metal). The sun was in the center of this whirling nebula which gradually flattened and developed into whirlpools which collected gas and dust in their centers. These collections of material eventually became the spinning planets that now travel around the sun.

Geologists have studied the rock layers of the earth and can trace its history. The age of rocks can be determined by measuring the amount of radioactive isotopes in them. The oldest rocks discovered on earth thus far are 4.3 billion years old. It is generally believed that all the planets in our solar system began to evolve approximately 4.5 billion years ago.

Scientists theorize that our planet slowly cooled from a gaseous state to molten rock, ultimately developing an outer crust of rock. The earth's interior was melted by heat from radioactive materials in the rock and the pressure from its mass. Heavy materials, such as iron, sank into the center of the earth. Lighter elements rose to form the earth's crust. This heating also caused other chemicals inside the earth to rise to the surface to form water and atmosphere.

The Scientific Method

The scientific method is a creative and systematic process for proving or disproving a given question following an observation. When scientists use the scientific method, a basic set of guiding principles and procedures is followed in order to obtain new knowledge about our universe. This method will be described in the paragraphs that follow.

It is easy to teach the scientific method! Just follow these simple steps:

 Make an **OBSERVATION**.

The teacher presents a situation, gives a demonstration, or reads background material that interests students and prompts them to ask questions. Or students can make observations and generate questions on their own as they study a topic.

Example: Show a sample of dirt from outside.

 Select a **QUESTION** to investigate.

In order for students to select a question for a scientific investigation, they will have to consider the materials they have or can get, as well as the resources (books, magazines, people, etc.) actually available to them. You can help them make an inventory of their materials and resources, either individually or as a group.

Tell students that in order to successfully investigate the questions they have selected, they must be very clear about what they are asking. Discuss effective questions with your students. Depending upon their level, simplify the question or make it more specific.

Example: What does dirt look like?

 Make a **PREDICTION** (*Hypothesis*).

Explain to students that a hypothesis is a good guess about what the answer to a question will probably be. But they do not want to make just any arbitrary guess. Encourage students to predict what they think will happen and why.

In order to formulate a hypothesis, students may have to gather more information through research.

Have students practice making hypotheses with questions you give them. Tell them to pretend they have already done their research. You want them to write each hypothesis so it follows these rules:

1. It is to the point.
2. It tells what will happen, based on what the question asks.
3. It follows the subject/verb relationship of the question.

Example: I think there are a bunch of living things in the dirt.

The Scientific Method *(cont.)*

Develop a **PROCEDURE** to test the hypothesis.

The first thing students must do in developing a procedure (the test plan) is to determine the materials they will need.

They must state exactly what needs to be done in step-by-step order. If they do not place their directions in the right order, or if they leave out a step, it becomes difficult for someone else to follow their directions. A scientist never knows when other scientists will want to try the same experiment to see if they end up with the same results!

Example: By examining several different samples of dirt, you will see many different living things.

Record the **RESULTS** of the investigation in written and picture form.

The results (data collected) of a scientific investigation are usually expressed two ways—in written form and in picture form. Both are summary statements. The written form reports the results with words. The picture form (often a chart or graph) reports the results so the information can be understood at a glance.

Example: The results of the investigation can be recorded on data-capture sheets provided (pages 14-15).

State a **CONCLUSION** that tells what the results of the investigation mean.

The conclusion is a statement which tells the outcome of the investigation. It is drawn after the student has studied the results of the experiment, and it interprets the results in relation to the stated hypothesis. A conclusion statement may read something like either of the following: "The results show that the hypothesis is supported," or "The results show that the hypothesis is *not* supported." Then restate the hypothesis if it was supported or revise it if it was not supported.

Example: The hypothesis that stated "there are a bunch of living things in the dirt" is supported (or not supported).

Record **QUESTIONS, OBSERVATIONS,** and **SUGGESTIONS** for future investigations.

Students should be encouraged to reflect on the investigations that they complete. These reflections, like those of professional scientists, may produce questions that will lead to further investigations.

Example: How is it possible for things to live in the dirt?

Science-Process Skills

Even the youngest students blossom in their ability to make sense out of their world and succeed in scientific investigations when they learn and use the science-process skills. These are the tools that help children think and act like professional scientists.

The first five process skills on the list below are the ones that should be emphasized with young children, but all of the skills will be utilized by anyone who is involved in scientific study.

Observing

It is through the process of observation that all information is acquired. That makes this skill the most fundamental of all the process skills. Children have been making observations all their lives, but they need to be made aware of how they can use their senses and prior knowledge to gain as much information as possible from each experience. Teachers can develop this skill in children by asking questions and making statements that encourage precise observations.

Communicating

Humans have developed the ability to use language and symbols which allow them to communicate not only in the "here and now" but over time and space as well. The accumulation of knowledge in science, as in other fields, is due to this process skill. Even young children should be able to understand the importance of researching others' communications about science and the importance of communicating their own findings in ways that are understandable and useful to others. The rocks and minerals journal and the data-capture sheets used in this book are two ways to develop this skill.

Comparing

Once observation skills are heightened, students should begin to notice the relationships between things that they are observing. Comparing means noticing the similarities and differences. By asking how things are alike and different or which is smaller or larger, teachers will encourage children to develop their comparison skills.

Ordering

Other relationships that students should be encouraged to observe are the linear patterns of seriation (order along a continuum: e.g., rough to smooth, large to small, bright to dim, few to many) and sequence (order along a time line or cycle). By making graphs, time lines, cyclical and sequence drawings, and by putting many objects in order by a variety of properties, students will grow in their abilities to make precise observations about the order of nature.

Categorizing

When students group or classify objects or events according to logical rationale, they are using the process skill of categorizing. Students begin to use this skill when they group by a single property such as color. As they develop this skill, they will be attending to multiple properties in order to make categorizations; the animal classification system, for example, is one system students can categorize.

Science-Process Skills *(cont.)*

Relating

Relating, which is one of the higher-level process skills, requires student scientists to notice how objects and phenomena interact with one another and the changes caused by these interactions. An obvious example of this is the study of chemical reactions.

Inferring

Not all phenomena are directly observable, because they are out of humankind's reach in terms of time, scale, and space. Some scientific knowledge must be logically inferred based on the data that is available. Much of the work of paleontologists, astronomers, and those studying the structure of matter is done by inference.

Applying

Even very young, budding scientists should begin to understand that people have used scientific knowledge in practical ways to change and improve the way we live. It is at this application level that science becomes meaningful for many students.

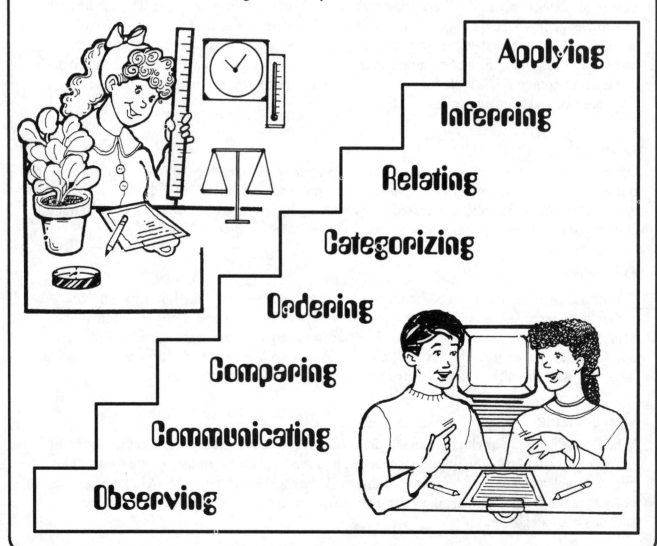

Organizing Your Unit

Designing a Science Lesson

In addition to the lessons presented in this unit, you will want to add lessons of your own, lessons that reflect the unique environment in which you live, as well as the interests of your students. When designing new lessons or revising old ones, try to include the following elements in your planning:

Question

Pose a question to your students that will guide them in the direction of the experience you wish to perform. Encourage all answers, but you want to lead the students towards the experiment you are going to be doing. Remember, there must be an observation before there can be a question. (Refer to The Scientific Method, pages 5-6.)

Setting the Stage

Prepare your students for the lesson. Brainstorm to find out what students already know. Have children review books to discover what is already known about the subject. Invite them to share what they have learned.

Materials Needed for Each Group or Individual

List the materials each group or individual will need for the investigation. Include a data-capture sheet when appropriate.

Procedure

Make sure students know the steps to take to complete the activity. Whenever possible, ask them to determine the procedure. Make use of assigned roles in group work. Create (or have your students create) a data-capture sheet. Ask yourself, "How will my students record and report what they have discovered? Will they tally, measure, draw, or make a checklist? Will they make a graph? Will they need to preserve specimens?" Let students record results orally, using a video or audio tape recorder. For written recording, encourage students to use a variety of paper supplies such as poster board or index cards. It is also important for students to keep a journal of their investigation activities. Journals can be made of lined and unlined paper. Students can design their own covers. The pages can be stapled or be put together with brads or spiral binding.

Extensions

Continue the success of the lesson. Consider which related skills or information you can tie into the lesson, like math, language arts skills, or something being learned in social studies. Make curriculum connections frequently and involve the students in making these connections. Extend the activity, whenever possible, to home investigations.

Closure

Encourage students to think about what they have learned and how the information connects to their own lives. Prepare journals using "Rocks and Minerals Journal" directions on page 79. Provide an ample supply of blank and lined pages for students to use as they complete the "Closure" activities. Allow time for students to record their thoughts and pictures in their journals.

Organizing Your Unit *(cont.)*

Structuring Student Groups for Scientific Investigations

Using cooperative learning strategies in conjunction with hands-on and discovery learning methods will benefit all of the students taking part in the investigation.

Cooperative Learning Strategies

1. In cooperative learning all group members need to work together to accomplish the task.

2. Cooperative learning groups should be heterogeneous.

3. Cooperative learning activities need to be designed so that each student contributes to the group and individual group members can be assessed on their performance.

4. Cooperative learning teams need to know the social as well as the academic objectives of a lesson.

Cooperative Learning Groups

Groups can be determined many ways for the scientific investigations in your class. Here is one way of forming groups that has proven to be successful in intermediate classrooms.

- **Expedition Leader**—scientist in charge of reading directions and setting up equipment
- **Geologist**—scientist in charge of carrying out directions (can be more than one student)
- **Stenographer**—scientist in charge of recording all the information
- **Transcriber**—scientist who translates notes and communicates findings

If the groups remain the same for more than one investigation, require each group to vary the people chosen for each job. All group members should get a chance to try each job at least once.

Using Centers for Scientific Investigations

Set up stations for each investigation. To accommodate several groups at a time, stations may be duplicated for the same investigation. Each station should contain directions for the activity, all necessary materials (or a list of materials for investigators to gather), a list of words (a word bank) which students may need for writing and speaking about the experience, and any data-capture sheets or needed materials for recording and reporting data and findings.

Station-to-Station Activities are on pages 73-77. Model and demonstrate each of the activities for the whole group. Have directions at each station. During the modeling session, have a student read the directions aloud while the teacher carries out the activity. When all the students understand what they must do, let small groups conduct the investigations at the centers. You may wish to have a few groups working at the centers, while others are occupied with other activities. In this case, you will want to set up a rotation schedule so all groups have a chance to work at the centers.

Assign each team to a station, and after they complete the task described, help them rotate in a clockwise order to the other stations. If some groups finish earlier than others, be prepared with another unit-related activity to keep students focused on main concepts.

After all rotations have been made by all groups, come together as a class to discuss what was learned.

Just the Facts *(Teacher)*

Rocks cover the entire surface of the earth, even beneath every body of water and the polar ice caps. This rock covering is referred to as the *crust* of the earth. Dirt or soil, which consists of crushed rock and pieces of organic material, covers some areas of the crust.

The earth's crust consists of three types of rock—*igneous, sedimentary,* and *metamorphic.* The crust is slowly and continuously recycled from one type of rock to another.

Igneous: Melted rock beneath the crust, called *magma,* is under tremendous pressure and sometimes rises through cracks in the crust. When magma solidifies, it is called igneous rock. Magma may cool underground within the crust or break through creating volcanoes which pour forth lava (magma above ground), which cools into igneous rock. The crust is cracked into large sections called plates. The edges of some crystal plates are forced beneath others, melting and recycling the leading edge of rock as it comes in contact with the hot magma.

Sedimentary: Sedimentary rocks consist of rocks which once were igneous, metamorphic, sedimentary or organic material. These materials are deposited in layers by wind, water, or ice. As the layers build up, the pressure packs the material together and squeezes out most of the water, forming solid rock layers. These sediments may consist of rock fragments ranging in size from large boulders to fine grains of sand and silt. These rocks may also be deposits of minerals in the form of crystals or organic sediments such as shells, skeletons, and plants. Fossils are found in sedimentary rocks.

Metamorphic: Metamorphic rock is igneous, sedimentary, or metamorphic rock which is subjected to tremendous pressure and heat by movement of the earth's crust or contact with magma. The original rock changes in appearance and often in mineral composition. For example, granite (an igneous rock), becomes gneiss, and calcite in limestone (a sedimentary rock), changes to marble.

The activities in this section will enable the students to develop an understanding of the rock cycle. They will analyze dirt and sand, as well as "walk through" the rock cycle.

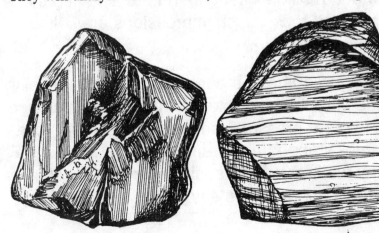

Igneous **Sedimentary**

Metamorphic

Just the Facts *(Students)*

The whole earth is covered by rocks. Over the ocean floors, along the river bottoms, under the ice and frozen snows, we find rocks. Even the dirt or soil is mainly just crushed rock. This rock covering for the earth is what we call the *crust.*

The earth's crust is made of three types of rock—*igneous, sedimentary,* and *metamorphic.* Each of these types of rock can be changed into another type. Here is how that can happen.

Igneous Rock: Miles deep under its crust, the earth is very hot—hot enough to melt rock and sand. This underground melted rock is called *magma.* When this magma cools and hardens, it is called igneous rock. The lava that oozes or spits out of a volcano is really just magma above ground. So lava that cools and hardens is igneous rock.

Sedimentary Rock: Over many years, rocks on the crust's surface will get worn away by wind and ice. Big rocks will finally become small. Pieces will become particles of dirt and grains of sand. These small particles may get carried away by wind and rain. They will be spread in layers along the land, river banks, and ocean floors. As more layers drift down on top of one another, the weight presses down more heavily. Over thousands of years this weight will pack the small grains of dirt and sand into solid layers of rock once again. This newly formed material we call sedimentary rock. Sometimes you can even find the skeletons of dead animals caught and pressed into these hardened layers of rock. Such impressions are called *fossils.*

Metamorphic Rock: Sometimes large parts of the earth's crust move, causing giant portions of rock and dirt to fold and press in on one another. This great pressure (and sometimes heat) can change the rock's appearance and make-up. It no longer resembles its previous sedimentary or igneous form. When this happens, we say the rock has become metamorphic.

An Ant's Eye View of Dirt

Question

What does dirt look like?

Setting the Stage

- A day or two before doing this activity, ask students to bring two tablespoons (30 g) of dirt from their own yards. Give each student a ziplock baggy with their names on them for their specimens.
- Discuss what dirt (soil) is with students.
- Tell students they will be collecting and examining dirt samples from the school yard.

Materials Needed for Each Individual

- small ziplock baggy
- magnifying lens
- data-capture sheets (pages 14-15)
  Note to the teacher: Samples of dirt from school and students' homes will be collected and examined to see what living and nonliving things are found. This activity is related to the one which follows: "Let's Eat Dirt."

Procedure

1. Divide students into groups of four.
2. Take students into the school yard to find different types of dirt samples. Look for a variety of areas to dig up samples, such as a grass area, a field of native plants, and a sand box.
3. Dig into the ground carefully to avoid damaging the site. Show students the nonliving and living things found in each sample.
4. Have students place samples of dirt from each area in separate ziplock baggies. On the baggy have them mark the location of the area from which the sample was taken.
5. Have students take the samples back into the classroom to examine with magnifying lenses. Have them record the information on their data-capture sheets.
6. Have students discuss with their group members the types of living and nonliving materials found in the dirt.

Extensions

- Have a class discussion about the most interesting living and nonliving materials they found in the dirt.
- Have students do research on different soil types. Then report their findings back to the class.

Closure

In their rocks and minerals journals, have students write a story and include drawings to describe how they think dirt is made, what lives in it, and where it is found.

An Ant's Eye View of Dirt *(cont.)*

Fill in the information needed below.

Name: _____

Date: _____

Dirt Specimen #1: _____

Location: _____

Color:_____

Smell: _____

Feel: _____

Living Things:

_____

_____

Nonliving Things:

_____

_____

Sample #1

Name: _____

Date: _____

Dirt Specimen #2: _____

Location: _____

Color:_____

Smell: _____

Feel: _____

Living Things:

_____

_____

Nonliving Things:

_____

_____

Sample #2

An Ant's Eye View of Dirt *(cont.)*

Fill in the information needed below.

Name: _____

Date: _____

Dirt Specimen #3: _____

Location: _____

Color: _____

Smell: _____

Feel: _____

Living Things:

_____

_____

Nonliving Things:

_____

_____

Sample #3

Name: _____

Date: _____

Dirt Specimen #4: _____

Location: _____

Color: _____

Smell: _____

Feel: _____

Living Things:

_____

_____

Nonliving Things:

_____

_____

Sample #4

Let's Eat Dirt

Question

How is dirt made?

Setting the Stage

- Have students work in small groups to read some descriptions from their Ant's Eye View of Dirt data-capture sheets.
- Ask students if they ever ate dirt when they were younger.
- Tell students that they are going to "make" dirt today and they will get to eat it.

Materials Needed for Each Group

- chocolate cookies, two per student
- 1 cup (250 g) powdered sugar
- 4 cups (1000 mL) of milk
- 2 cups (500 g) of raisins
- 8 oz. (250 g) softened package of cream cheese
- ½ cup (125 g) softened margarine or butter
- jellied sugar worms, one per student
- chocolate sprinkles
- two 4 oz. (120 g) packages of instant chocolate pudding
- 8 oz. (250 g) container of whipped topping
- clear plastic drinking straw, one per student
- clear plastic 6 oz. (180 mL) cup
- sealable plastic bag, one per student
  Note to the teacher: Layers of dirt are simulated in an edible form to illustrate what dirt contains.

Procedure

1. Have all students wash their hands thoroughly before beginning this activity.
2. Distribute to each student one plastic bag and two chocolate cookies. Tell them to place the cookies inside the plastic bag, press out the air, and then zip it closed.
3. Have each student place a book over the plastic bag and press hard until the cookies are crushed into small pieces. Tell them the cookies represent rocks being broken into smaller pieces to make dirt.
4. Divide students into groups to assist in preparing the ingredients as follows:
   - Cream together the softened cream cheese, butter, and powdered sugar.
   - Fold the whipped topping into the creamed mixture.
   - Beat the milk into the instant pudding.
5. Distribute small bowls of the creamed mixture and pudding mixture to each group.

Let's Eat Dirt *(cont.)*

6. Give each group raisins, jellied worms, and chocolate sprinkles. Explain that the raisins represent small rocks, the jellied worms are earthworms, and the chocolate sprinkles are small insects like ants.

7. Give each student a plastic cup and have them place a thin layer of crushed cookies at the bottom.

8. Then, have them begin spooning layers of the creamed and pudding mixtures alternately in their cups. Also, continue to add layers of crushed cookies. Between each layer they should also add their raisins, worms, and sprinkles. Finally, put a layer of crushed cookies on top.

9. Chill the cups of "dirt" until they are set.

Extensions

Give each student a clear plastic straw and tell them to push it straight down into the dirt and then pull it out again. Ask them to draw this core sample of dirt. Have students go to a place where layers of dirt are visible (e.g., road cut or hillside) and compare it with their core sample.

Closure

As a finale, let the students eat their homemade "dirt" samples—rocks, insects, and all.

Teeny Tiny Rocks

Question

What is sand?

Setting the Stage

- Ask students how they think sand is made. (Do not tell them the answer.)
- Tell students they will discover where sand comes from by participating in the next experience.
- Ask for student help in collecting at least four different sand samples.

Materials Needed for Each Group

- magnifying lens
- transparent tape
- scissors
- sand samples (already collected)
- data-capture sheets (pages 19-20), one per student

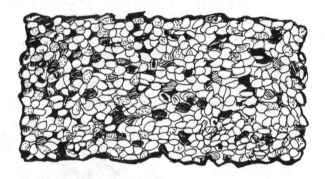

Note to the teacher: Sand samples will be examined to discover that sand is made of a variety of crushed minerals.

Procedure

1. Make sets of four sand samples in open dishes and label them A through D.
2. Divide students into groups and provide each group with a set of sand samples.
3. Distribute all materials to the groups.
4. Have students cut out the holes shown on their data-capture sheets where they are to tape their sand samples A through D.
5. Demonstrate how students should put a piece of transparent tape behind the holes they cut out (sticky side toward the front of the paper) and then press the sticky side into the correct sample of sand A through D.
6. After students have the four sand samples taped to their papers, tell them to fill in the information needed on their data-capture sheets.

Extensions

- Demonstrate for your students how sand is made by pulverizing a small rock. This should be done while wearing safety goggles, placing the rock between layers of newspaper, and then hammering so the chips will not fly out and hit anyone.
- If possible, take students to a nearby beach or stream so the students can collect their own sand samples and observe how the sand is being formed by water action.

Closure

In their rocks and minerals journals, have students write about what they have learned. Then they can share their new knowledge with the class.

Teeny Tiny Rocks *(cont.)*

Fill in the information needed below.

Sand Specimen A

Colors: _____

Feel: _____

Cut out the rectangle below and tape a sample of Sand A there.

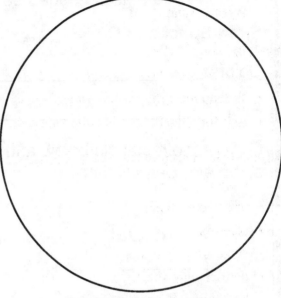

Cut Out

Drawing of sample A
through a magnifier:

Sand Specimen B

Colors:_____

Feel: _____

Cut out the rectangle below and tape a sample of Sand B there.

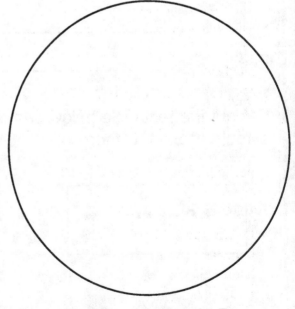

Cut Out

Drawing of sample B
through a magnifier:

Teeny Tiny Rocks *(cont.)*

Fill in the information needed below.

Sand Specimen C

Colors:_____

Feel: _____

Cut out the rectangle below and tape a sample of Sand C there.

Cut Out

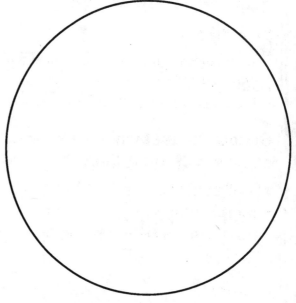

Drawing of sample C through a magnifier:

Sand Specimen D

Colors:_____

Feel: _____

Cut out the rectangle below and tape a sample of Sand D there.

Cut Out

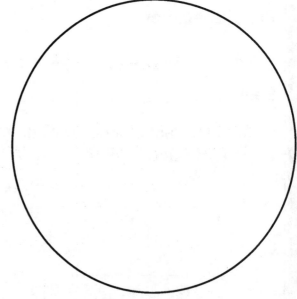

Drawing of sample D through a magnifier:

The Crust of The Earth

Question

How much of the earth is made of rock?

Setting the Stage

- Have students draw what they think the earth would look like if it were cut in half. Have them draw the surface as well as the inside of the earth.
- Have students share their drawings with the class.

Materials Needed for Each Group

- reproducible of earth's layers (page 22), one per student
- hardboiled egg
- permanent black marker

Note to the teacher: The earth's layers will be explained through the use of an egg and a diagram.

Procedure

1. Cut a hard boiled egg in half lengthwise and use a permanent black felt pen to make a dot in the center of the egg.

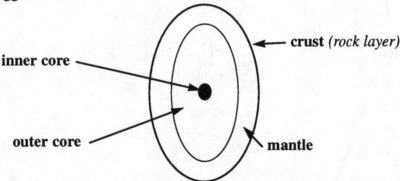

2. Show a picture of the earth's layers and describe the different layers to the class. Explain that the earth's diameter is about 8,000 miles (12,800 km). The crust of the earth is like the egg shell, only it is thinner relative to the earth than the shell is to the egg. The white of the egg is like a mantle which is extremely hot inside the earth. The yolk represents the outer core, and the black dot is the inner core. Point out that the deeper you go inside the earth, the hotter it becomes. This is due to both the pressure of the earth towards the center, and the crust preventing the inside of the earth from ever cooling. The earth is like a pie: the crust cools before the filling does.

Extensions

Have students compare their drawings of the earth with page 22 before they participate in this experience.

Closure

In their rocks and minerals journals, have students draw and color a new diagram of the earth and all of its layers.

The Crust of the Earth *(cont.)*

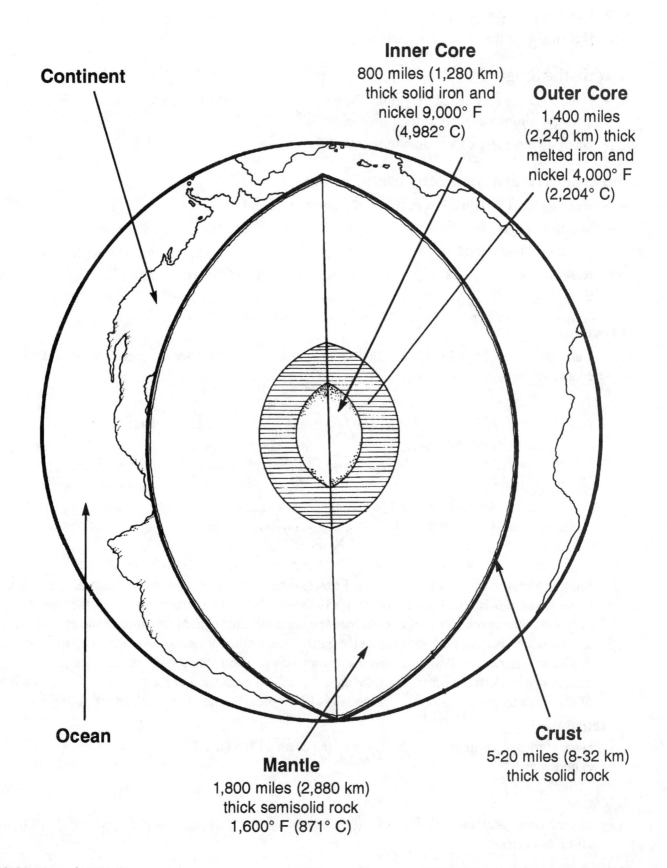

Continent

Inner Core
800 miles (1,280 km) thick solid iron and nickel 9,000° F (4,982° C)

Outer Core
1,400 miles (2,240 km) thick melted iron and nickel 4,000° F (2,204° C)

Ocean

Mantle
1,800 miles (2,880 km) thick semisolid rock 1,600° F (871° C)

Crust
5-20 miles (8-32 km) thick solid rock

Rocks Go Round and Round

Question

What is the rock cycle?

Setting the Stage

- Make a transparency of The Rock Cycle (page 24) and use it to explain the rock cycle to students.
- Explain to students that rocks are either igneous, metamorphic, or sedimentary and are constantly being recycled.

Materials Needed for Each Group

- large piece of butcher paper or a bed sheet
- colored markers or crayons
- copies of Rocks Go Round and Round (pages 24-25), one each per student

Note to the teacher: The rock cycle will be simulated in a walk through the earth's crust.

Procedure

1. Draw onto your butcher paper or bed sheet an enlarged picture of Rocks Go Round and Round (page 25). Color the igneous rock black, the mantle and lava red, sedimentary rock brown, metamorphic rock various colors, and the ocean water blue. (You may wish to have students help here.)

2. Lay the enlarged drawing on the floor, permitting space for students to walk on it.

3. Demonstrate the walk through the rock cycle while reading aloud the following:

   - This journey begins in the mantle beneath the earth's crust.
   - Magma is forced up, changing to lava and spreading between cracks in a volcano, eventually cooling to become igneous rock. Some of the lava pours out the top and down the sides of the volcano.
   - The lava rolls down the sides of the volcano to the ocean, cooling along the way or when it flows into the water and changes into igneous rock.
   - Wave action breaks the lava and igneous rock into sand-sized pieces.
   - More layers of the sand are added and pressed together, becoming sedimentary rock.
   - More layers of sedimentary rock are formed on top of that layer. Some sedimentary rocks on the bottom get hot because of the pressure and change to metamorphic rock.
   - When the metamorphic rock is buried even deeper, it gets hotter and melts, becoming magma. It may eventually be pushed up again into the crust.

4. Explain to students that this is just one way the rocks in the earth's crust are recycled, for rocks are constantly changing. The rock cycle happens very slowly over millions of years.

5. Have students from each group walk through the rock cycle as you read about their journey.

Extensions

Have students collect rocks from the three categories—igneous, metamorphic, and sedimentary—and make a rock cycle on cardboard, showing the direction the cycle moves.

Closure

Have students color in their copy of Rocks Go Round and Round and add it to their rocks and minerals journal.

Rocks Go Round and Round *(cont.)*

The Rock Cycle

Igneous

Rock formed from melted igneous, sedimentary, or metamorphic rock.

Sediment

Broken pieces of shell, sedimentary, igneous, or metamorphic rock, and parts of plants and animals.

Metamorphic

Igneous, sedimentary, or metamorphic rock which has been changed by heat and pressure.

Sedimentary

Rock made from compressed pieces of sediment.

Examples of the Three Types of Rocks

Igneous—obsidian, pumice, and granite

Metamorphic—marble (from limestone) and slate (from shale)

Sedimentary—sandstone, limestone, and shale

Note: Quartz is found in all three of these types of rocks.

Rocks Go Round and Round *(cont.)*

Recycling the Earth's Crust

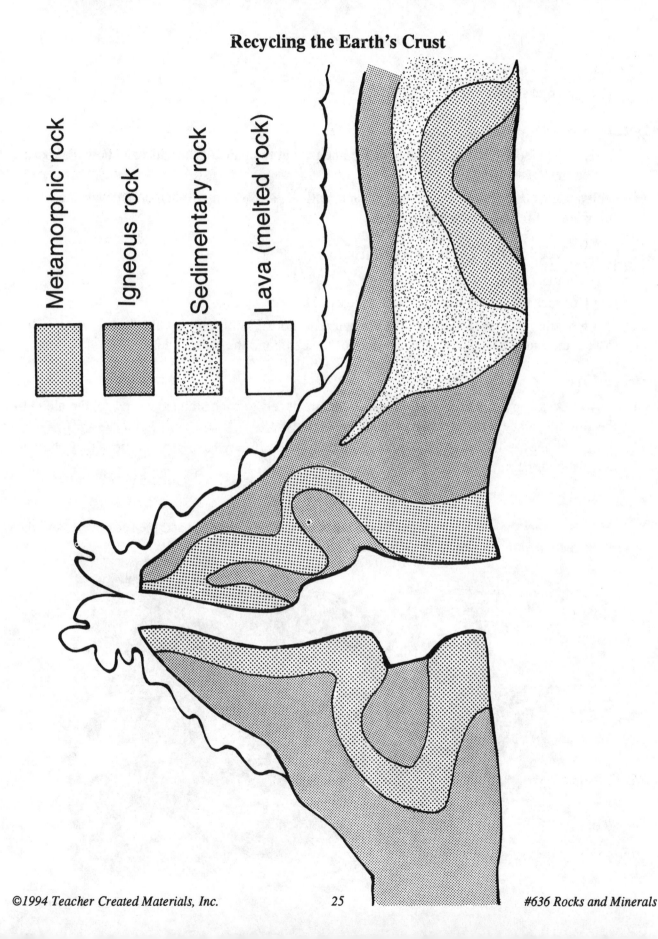

Metamorphic rock

Igneous rock

Sedimentary rock

Lava (melted rock)

The Rock Cycle

Assessment Activity

Question

How are rocks recycled?

Setting the Stage

- Ask students to think of what they have learned about the parts of the earth and how the rocks in the earth's crust change.
- Tell students they will be completing a data-capture sheet which will help you determine what they learned from the previous activities.

Materials Needed for Each Individual

- colored markers or crayons
- copy of rock cycle data-capture sheets (pages 27-28)

Note to the teacher: Students will complete The Rock Cycle data-capture sheets, which will assess their understanding of the structure of the earth and the rock cycle.

Procedure

1. Distribute a copy of The Rock Cycle data-capture sheets to each student and read them aloud to be sure they understand how to complete the activity.
2. Tell them they will be doing this activity on their own. If they have any questions they should ask for your help.

Closure

Have students share their drawings of the cutaway view of the earth and the rock cycle, and then have them add the drawings to their rocks and minerals journal.

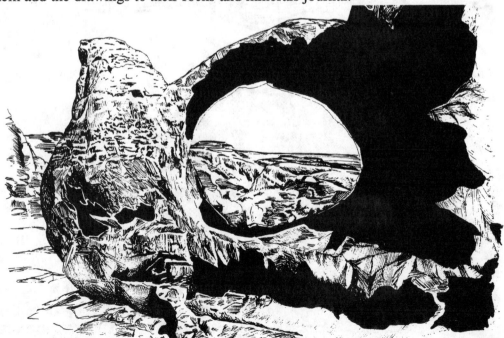

The Rock Cycle *(cont.)*

Draw and complete the pictures below.

The circle represents the earth cut in half.

Finish this picture by following the key to color the parts of the earth:

Key   **the earth**

Part of the earth   ***Color***

   crust   brown

   mantle   red

   outer core   yellow

   inner core   black

Draw a picture below to explain the rock cycle.

Put labels on your drawing to tell the different parts.

Color your drawing to show the three types of rocks.

The Rock Cycle

The Rock Cycle *(cont.)*

Explain the rock cycle you have shown in your drawing:

_____

_____

_____

_____

_____

_____

_____

_____

_____

_____

_____

_____

_____

_____

_____

_____

Just the Facts *(Teacher)*

Rocks are made up of two or more minerals. Minerals are found in soil and are also found on the moon, Mercury, Venus, and Mars. They are used to make many products—for example, cement, fertilizers, and chemicals for manufacturing. Common minerals such as graphite are used in "lead" pencils, and rare minerals such as gold and silver are used for jewelry and money.

Mineralogists (scientists who study minerals) define minerals as substances which:

> formed naturally.
>
> are made of materials that were never alive.
>
> have the same chemical makeup wherever they are found.
>
> have atoms which are arranged in regular patterns and form solid units called crystals.

By this definition substances such as coal, petroleum and natural gas, or pearls and coral, are not minerals since they were formed by once living plants and animals. Substances such as calcium, iron, and phosphorous, which are found in food and water, are often referred to as minerals, but mineralogists do not consider them minerals.

Minerals are usually a compound of two or more elements. Some minerals, such as gold and sulfur, however, are made of only one element. The most common elements which form minerals are oxygen and silicon. Others include aluminum, iron, calcium, sodium, potassium, and magnesium. There are about 2,000 different minerals known. Common minerals can be recognized by examining some of their characteristics such as the following:

Color: Minerals are found in a variety of colors due to the chemicals in them. For instance, quartz occurs in many hues but may also be colorless. Some minerals are always the same color—e.g., galena is metallic gray, sulfur is yellow, azurite is blue, and malachite is green. A fresh surface is needed to see the true color since weathering may hide it.

Luster: The amount of light reflected from the mineral's surface is its luster. Luster may be described as glassy, metallic, shiny, dull, waxy, satiny, or greasy.

Streak Color: Some minerals leave a colored streak when rubbed across a piece of unglazed white tile. The streak color may not be the same as the mineral's color. For example, hematite may be black to brown, but its streak is red brown.

Texture: Texture is the "feel" of the mineral's surface when it is rubbed. This may be rough, smooth, bumpy, or soapy.

Hardness: Although all minerals are hard, the surface varies in resistance to scratching. Mohs' hardness scale of 1 to 10 is applied to minerals. The hardness test is done with common materials which vary in hardness, such as a fingernail, penny, steel knife or nail, and glass. The hardness number is assigned depending upon which item will scratch the mineral's surface.

The activities in this section will develop the students' skills in identifying minerals through careful observation and by gathering data about the characteristics described above.

Just the Facts (Students)

All rocks are made up of two or more minerals. Minerals are found in the earth's crust and on some planets. They are used to make products such as cement and fertilizer. Some minerals are common, like graphite used in pencils for writing. Some minerals are rare, like gold and silver used for making money and jewelry.

Scientists who study minerals are called *mineralogists.* They tell us that minerals are formed naturally out of materials that were never alive. Things like coal or pearls are not minerals, since they are formed by once living things.

We can learn to recognize many common minerals by looking for these five qualities:

Color: Some minerals are always the same color. Galena is always metallic gray, sulfur is always yellow, azurite is always blue, and malachite is always green.

Luster: Different minerals reflect light in different ways. This reflection or luster is described as glassy, metallic, shiny, dull, waxy, satiny, or greasy.

Streak Color: Some minerals leave a colored streak when rubbed across a piece of unglazed white tile. The streak color may not be the same as the mineral's color. The mineral hematite may be black to brown, but its streak color is red brown.

Texture: The feel of a mineral's surface when it is rubbed is called texture. The texture can be rough, smooth, bumpy, or soapy.

Hardness: Some minerals are harder than others and not so easy to scratch. We can check the hardness by trying to scratch the mineral with something soft like a fingernail. If that does not work, we try a penny. If that will not scratch the mineral, we try glass. Finally, if the glass will not scratch it, we try a steel nail.

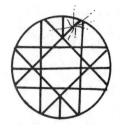 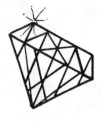 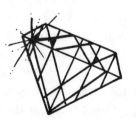 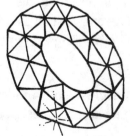 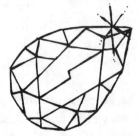

The Sorting Game

Question

How can things be sorted?

Setting the Stage

- Tell students you are going to sort them into groups and that you want them to see if they can guess how you are sorting them.

- Sort students into two groups by sex, without letting the students know your system. Do this by calling one student at a time to go to a specific area of the room so they will join the boys' or girls' group.

- After you have selected about five or six students, see if students can guess your system of sorting by asking a few students to join the group to which they think they belong.

- Explain to students that there are a variety of ways to sort things and that they will do an activity to demonstrate this.

Materials Needed for Each Individual

- one of their shoes
- data-capture sheet (page 32)

  Note to the teacher: The students will use their shoes to learn how to sort them according to various properties.

Procedure

1. Seat students in a large circle on the floor and ask them to remove one shoe (You should also add one of your shoes.)
2. Place the shoes side by side and ask students to suggest a way they can be sorted using only one property at a time (e.g., color, fabric, size).
3. Have two students sort the shoes into piles according to the property chosen by the class.
4. Now have students choose another property by which to sort the shoes.
5. Have students record on their data-capture sheets, the different properties by which the shoes were sorted. Then, mark the number of shoes that possessed each property.

Extensions

Have students continue this activity until they know that they can only use one property at a time for sorting the shoes but that there are many different properties from which to choose.

Closure

Divide students into groups of three or four and give each group a deck of cards from which the jokers have been removed. Tell students to sort the cards. Have them repeat the sorting using other properties. Discuss the variety of properties used by each group to do the sorting.

The Sorting Game *(cont.)*

Shoe Sorting Properties	Number of Shoes with Each Property

Did any shoes not fall under any of the above classifications? Explain.

_____

_____

_____

Mineral Mystery

Question

How can you describe minerals?

Setting the Stage

- Display rock and mineral specimens and pictures in the classroom. Ask students to bring rocks and minerals they may have collected to add to this display.
- Discuss what the students learned by sorting their shoes in the previous lesson.
- Tell students that today they will look at eight mineral samples and sort them by different properties.

Materials Needed for Each Group

- set of 8 mineral specimens: calcite, galena, graphite, hematite, magnetite, obsidian, quartz, and talc (See page 96 for resources.)
- data-capture sheet (page 34), one per student

   Note to the teacher: Students will discover a variety of properties which can be used to identify minerals.

Procedure

1. Divide students into groups of three or four and distribute a set of eight minerals to each group.
2. Tell students to look at their minerals and find one property which they can use to sort them into separate piles.
3. Explain that they need to leave these minerals in their separate piles since after they finish, they will move to another group's table and try to find out what property they used.
4. Let the students sort their minerals. Monitor their progress to be certain they choose only one property and that it is one which relates to the mineral specimens.
5. After the groups have sorted their minerals, have them rotate to another group's area and decide what property they used to sort them. Have them check with the group who did the sorting to be certain they are correct.
6. Let each group sort their new set of minerals using a different property this time.
7. Have students complete their data-capture sheets by listing the various properties used to sort the minerals. This list will be used in the next activity.

Extensions

Rotate the groups at least two times to encourage the students to look for less obvious properties.

Closure

In their rocks and minerals journals, have students write about all the different minerals they have discovered.

Mineral Mystery *(cont.)*

Make a list below of the properties used to sort your minerals.

1. _____

2. _____

3. _____

4. _____

5. _____

6. _____

7. _____

8. _____

9. _____

10. _____

Mineral Detectives

Question

How can minerals be identified?

Setting the Stage

- Show students the list of mineral properties which they discovered in the last lesson.
- Introduce to students some other properties which they may not have discovered but which are used to help identify the minerals:

    Hardness—measured on a scale of 1 - 10 and tested by scratching the mineral with a fingernail, penny, streak plate, and steel nail. (See Mohs' Hardness Scale on page 36.) These tools are used in the order shown. It is not necessary to scratch the mineral with all the tools, for the first to scratch the surface will determine the hardness. A magnifying lens may be helpful to see the surface more clearly to check for scratches.

    Streak Color—This test requires the use of an unglazed tile called a streak plate. Rub a corner of the mineral across the streak plate several times to see if it leaves a color. Some streak colors are different from the color of the mineral. If a mineral is harder than the streak plate (6.5 or greater), it will not leave a streak color. A magnifying lens may be used to examine the streak to find its exact color.

    Luster—surface appearance of a mineral, described as glassy, metallic, dull, or shiny.

    Texture—feel of the surface of a mineral, described as rough, smooth, soapy, or bumpy.

Materials Needed for Each Group

- set of eight mineral specimens
- magnifying lens
- hardness kit—penny, streak plate (unglazed tile), steel nail
- Mohs' Hardness Scale (page 36), one per student
- data-capture sheet transparency (page 37), teacher only
- data-capture sheet (page 37), one per student

    Note to the teacher: Students will learn how to identify the eight mineral specimens.

Procedure

1. Divide students into groups of three or four and distribute the materials they will use for this activity.
2. Demonstrate how to describe the minerals by selecting one and writing the description of its color, luster, texture, streak, and hardness on a transparency copy of the data-capture sheet.
3. Each group selects six of their eight minerals to describe and places them on the numbers.
4. Have each member of the group write the description of the minerals on their data-capture sheet.
5. Have students use their copy of Mohs' Hardness Scale to refer to as they work.

Extensions

Have students repeat this experience, this time using eight different mineral samples.

Closure

Have students complete their data-capture sheets and then add them to their rocks and minerals journals.

Mineral Detectives *(cont.)*

Mohs' Hardness Scale
from 1 (softest) to 10 (hardest)

Mineral Examples

Mineral	Hardness
Talc	1
Gypsum	2
Calcite	3
Fluorite	4
Apatite	5
Feldspar	6
Quartz	7
Topaz	8
Corundum	9
Diamond	10

Hardness Test

Minerals scratched by:	Are a hardness of:
fingernail	2.5 or less
penny	3 or less
glass (streak plate)	5.5 or less
steel nail	6.5 or less
none of the above	greater than 6.5

Mineral Detectives *(cont.)*

Complete the chart below.

Rock	Color	Luster	Texture	Streak Color	Hardness
1					What scratched the mineral? _____ Hardness number? _____
2					What scratched the mineral? _____ Hardness number? _____
3					What scratched the mineral? _____ Hardness number? _____
4					What scratched the mineral? _____ Hardness number? _____
5					What scratched the mineral? _____ Hardness number? _____
6					What scratched the mineral? _____ Hardness number? _____

Match the Minerals

Question
Can you match the mineral with its description?

Setting the Stage
- Divide students into the same groups as they were during the Mineral Detectives activity.
- Distribute the mineral specimens and data-capture sheets from the last activity to each group.
- Tell students to match their specimens with their descriptions. This review of the last activity will prepare them for the lesson they are about to do.
- Have students remove their minerals from their charts and rotate to another group's area where they will match their minerals to the other group's descriptions.

Materials Needed for Each Group
- set of eight mineral specimens
- Mineral Identification Key (page 39), one per student
- magnifying lens
- hardness kit—penny, streak plate (unglazed tile), steel nail
- data-capture sheet (page 40), one per student

    Note to the teacher: Students will use the minerals and their data-capture sheet completed in the last activity to reinforce their skills of identifying minerals.

Procedure
1. Divide students into groups of three or four and distribute the minerals and other materials.
2. Tell students to use the Mineral Identification Key to identify their minerals and place them on their data-capture sheets as they find their names.

Extensions
When all groups have finished identifying their minerals, let student groups check each other's work.

Closure
Have students sort the class rock collection and then see if they can find any of the eight minerals among the specimens. Provide simple mineral identification books for the students to try to identify the rocks in the collection.

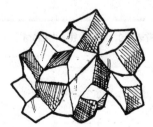 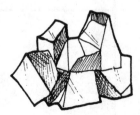  

Match the Minerals *(cont.)*

Mineral Identification Key

Match your minerals to these descriptions and then find their names.

Mineral	Color	Luster	Texture	Streak	Hardness
calcite	tan & white	shiny and glassy	smooth	white or pink	penny 3
galena	silver	metallic and shiny	smooth to rough	dark grey or black	penny 3
graphite	dark grey	dull	smooth to bumpy	black or dark grey	fingernail 1
quartz	milky white	shiny and glassy	smooth to bumpy	white	none more than 6.5
obsidian	black	glassy	smooth with sharp edges	none	none more than 6.5
hematite	reddish brown	dull	rough	red brown	fingernail to nail 1-6
magnetite	grey or black	dull	rough	black or dark grey	none above 7
talc	light grey may have some white	dull	smooth, feels like soap	white	fingernail 1

Match the Minerals *(cont.)*

Place your minerals on this chart after you have identified them.

Calcite	Galena
Graphite	**Hematite**
Magnetite	**Obsidian**
Quartz	**Talc**

Rock Cookies

Question

How can you tell a rock from a mineral?

Setting the Stage

- Let students examine the rocks from the class collection which have several different minerals in them.
- Show the eight mineral samples and let students compare the rocks and minerals.
- Explain to students that rocks are made of a mixture of two or more minerals.
- Tell students that minerals are made of one or more chemicals.

Materials Needed for Each Individual

- cookie which contains a variety of ingredients (e.g., raisins, chocolate chips, and nuts)
- paper towel
- crayons or colored markers
- data-capture sheet (page 42)

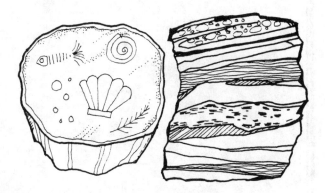

Note to the teacher: Students will use cookies to learn the difference between rocks and minerals.

Procedure

1. Explain to students that the cookies represent rocks and as they break each cookie apart, they should look for pieces which represent minerals (i.e. raisins, chocolate chips, and nuts).
2. Let the students separate the minerals into piles on the paper towel.
3. Have them compare their rock cookies with the natural rocks to see if they can find minerals in the rocks.
4. Then, have them draw their rock cookies on their data-capture sheets.
5. Finally, let them eat their rock cookies.

Extensions

Let students make rock cookies of their own so they can put together the minerals needed to create their rocks. They may add a variety of minerals such as chocolate chips, marshmallows, and nuts.

Closure

In their rocks and minerals journals, have students draw a rock with minerals in it and a single mineral. This will check for understanding of the concept that rocks are made of one or more minerals but a mineral is all the same material.

Rock Cookies *(cont.)*

Draw a picture of your rock cookie.

Homemade Rocks

Question

Can you make a rock?

Setting the Stage

- Show students a variety of rock samples and ask them to tell you what is the difference between rocks and minerals. (A rock is made of a variety of minerals.)
- Tell students that in this activity they will make a rock using play dough and things which they bring from home. Distribute the parent letter (page 44.)
- Have students take these home so their families can help find things to go into their rocks.

Materials Needed for Each Individual

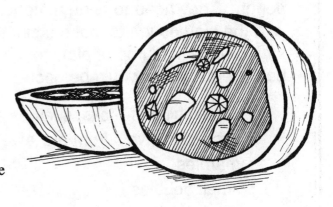

- copy of parent letter (page 44)
- play dough (about the size of a golf ball)
- materials to mix into the play dough (e.g., small chips of rocks, marbles, charcoal pieces, and scraps of aluminum)
- 10" (25 cm) of string

  Note to the teacher: Simulated rocks will be made using a variety of materials which are collected by students and teacher.

Procedure

1. Distribute a lump of play dough to each student and have them knead it until it is soft.
2. Tell students to slowly mix the play dough with the things they brought to represent minerals.
3. Each student should add another thin layer of play dough to their rocks. This represents the weathering which takes place in nature, often hiding the minerals which are inside the rock.

Extensions

Give each student a piece of string. Have them tie each end around a separate pencil so that about 4" (10 cm) of string is left between the pencils. Show them how to use this device to cut their rocks in half. Have everyone examine the cross section of their rocks and look at the minerals which they can now see that were not visible on the surface before.

Closure

Show students a variety of rocks which have been cracked open to demonstrate that sometimes the inside of a rock does not look much like the outer layer. This is especially true of geodes which may contain cavities filled with crystals. Have them compare their "rocks" with those which formed naturally.

Homemade Rocks *(cont.)*

Parent Letter for Homemade Rocks

Dear Parents,

We are studying rocks and minerals in our science class. Our next activity is to make rocks for which each child will receive some play dough. They need to bring materials from home which can be mixed with the play dough to make a simulated rock. These items should be small, since the lump of play dough will be about the size of a golf ball. Items which would be good to use for this activity are as follows:

- small chips of rocks, such as aquarium rocks
- synthetic jewels from old jewelry
- marbles
- small pebbles
- charcoal pieces
- scraps of aluminum foil

When the students are finished making their rocks, they will cut them open and compare them with real rocks which have been opened. They will be able to keep their rocks at the end of our study.

Sincerely,

Which Mineral Is This?

Assessment Activity

Question

How can you identify a mineral?

Preparation of Supply Centers

- Set out five streak plates and five hardness kits (penny, streak plate, nail) in separate areas around the room.
- Place a copy of Mohs' Hardness Scale and a magnifier in each hardness test center.
- Post copies of the Mineral Identification Key in different locations around the room.

Setting the Stage

- Ask students to think of the techniques they have learned to identify minerals.
- Tell students that in this activity they will receive one of the eight minerals to identify so that they can show what they have learned.

Materials Needed for Each Individual

- one of the eight minerals used in the previous activities
- data-capture sheet (page 46)

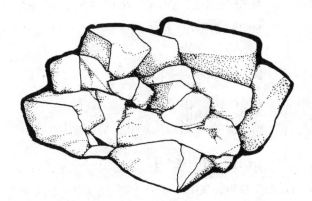

  Note to the teacher: Students will identify one of the eight mineral specimens, applying the techniques learned in the series of activities in this section.

Procedure

1. Distribute a copy of the data-capture sheet to each student and read it aloud to them to be sure all students know what they are expected to do.
2. Show the students the locations of the test centers for streak color and hardness, as well as where copies of the Mineral Identification Key are located in the room.
3. Tell students that as they work on the descriptions of their minerals they may not talk and that only one person at a time can work at a center. Explain that they do not need to work on the descriptions for their mineral in the order shown on their worksheet. If the centers are busy, they should complete a part of their worksheet that does not need any special equipment.
4. Let students begin. Monitor their progress and evaluate their skills as they work.

Closure

Have students read their descriptions aloud to see if others can guess their mineral. Identify all the minerals so students will know if they were right.

Which Mineral Is This? *(cont.)*

Fill in the information needed below.

Name: _____Date:_____

The color of my mineral is _____.

The texture (feel) of my mineral is _____.

My mineral has a _____luster.

When I rub my mineral across a streak plate, it leaves the color _____.

Circle the tool which scratched your mineral in the hardness test:

      fingernail

      penny

      glass (streak plate)

      steel (nail)

      none of these

Look at Mohs' Hardness Scale to find the hardness of your mineral.

The hardness of my mineral is _____.

Write any other details about your mineral which might help another person pick it out of a pile of minerals:

_____

_____

After you have finished everything above, look at the Mineral Identification Key and find the name of your mineral.

The name of my mineral is _____.

Just the Facts *(Teacher)*

The most attractive feature of minerals is their crystal form. If you could crawl inside a crystal, it would look like rooms formed by the crystal's atoms. For instance, a "room" in a copper crystal is formed by 14 copper atoms, one atom at each corner of the floor and ceiling, and an atom at the centers of the floors, the ceiling, and each of the four walls. The copper crystal consists of many of these rooms packed together. These rooms are called unit cells and may have four or six walls. Some have slanted walls. The shape of the crystal depends upon the shape of the unit cells. Crystals are classified according to their shape, such as salt which is isometric or cubic in shape.

Crystals may form from vapors, solutions, or melted materials. When temperature or pressure is lowered or evaporation takes place, certain atoms move closer together and join to form unit cells. A crystal grows larger as more atoms are added to its surfaces expanding the unit cells. The more slowly crystals take to form, the larger they become.

The most spectacular of mineral crystals are gems such as diamonds, rubies, or emeralds. Some minerals have crystals which are always one color, but other crystals may have several different colors for the same mineral. For example, quartz crystals may be milky white, clear, yellow, brown, or violet.

Sometimes crystals form inside cavities of rocks. When rock cavities are partly filled with crystals, they are called *geodes*. Opening a geode is like breaking into a treasure chest, for sometimes the crystals inside it are a beautiful color and may even be fairly large.

The activities in this section will lead the students through the excitement of growing crystals and recording the results.

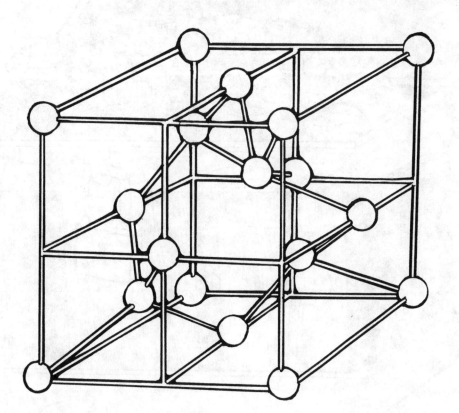

Just the Facts *(Students)*

Most minerals are formed in liquid and develop as crystals. That means they have solid, regular shapes. These shapes can be simple or complex, but they are always regular and beautiful. Sometimes these beautiful mineral crystals are so small that we need special instruments to see them. Sometimes the crystals are large and colorful.

Probably the most famous mineral crystals are gems such as diamonds, rubies, and emeralds. Some crystals are always the same color. Others may vary, like quartz which may appear clear, milky white, yellow, brown, or even violet.

A very special kind of crystal formation takes place inside other rocks. When such rocks are broken open, we will find a cavity filled with beautiful crystals. Rocks containing these cavities with crystals are called *geodes*. It is a great thrill to find a geode, for it is like finding a treasure chest. Perhaps you have seen one of these treasures displayed in a jewelry store. It may have contained the beautiful purple or violet mineral called *amethyst*.

Crystal Creations

Question

How do crystals form?

Setting the Stage

- Display samples or pictures of mineral crystals in the classroom.
- Explain to students that there are many different mineral crystals which vary in shape and color.
- Tell students that they are going to make crystals in this activity.

Materials Needed for Each Group

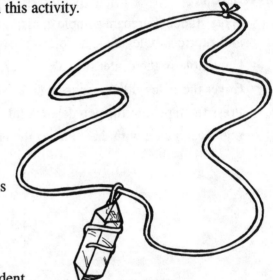

- shallow dish or petri dish
- clear disposable 5 oz. (148 mL) cups
- disposable spoon or stirring stick
- hot water
- * salt, alum, Epsom salt, cupric sulfate
- red, blue, and green food coloring in dropper bottles
- magnifying lenses
- scissors
- transparent tape
- data-capture sheets (pages 51-52), one each per student

    * Most of these chemicals can be found in the grocery or drugstore. Cupric sulfate is poisonous and thus should be demonstrated by the teacher. It can be obtained from a scientific supply company (See Resource section.) or high school chemistry teacher.

    Note to the teacher: Four different chemicals will be dissolved in water. The solutions will be poured into shallow dishes to evaporate, leaving the crystallized chemicals behind.

Teacher Demonstration

Demonstrate how to mix the crystal solution using the cupric sulfate as follows:

- Pour about 3 teaspoons (15 mL) of hot water (not boiling) into the clear cup.
- Add about a ¼ teaspoon (1 g) of cupric sulfate to the water and stir until it dissolves.
- If necessary, continue to add the sulfate until no more will dissolve.
- Pour all of this solution into the shallow dish, including any undissolved solids.

Procedure

1. Divide students into six groups and give each group a dish, spoon, and cup with 3 teaspoons (15 mL) of hot water in it.
2. Give each group 1 tablespoon (15 g) of salt, alum, or Epsom salt. There will be two groups using each of the chemicals.
3. Tell students to gradually add the chemical with the spoon, stirring after each addition until the solid has dissolved, and then continue adding it until no more will dissolve.

Crystal Creations *(cont.)*

4. Give a container of food coloring to one group working with salt, alum, and Epsom salt. Explain that this will be like the crystals found in nature which are colored by chemical impurities.

5. Have students pour their solutions into the dish, including any undissolved solids.

6. Let the solution sit for several days until all the water is evaporated.

7. Have students tape a sample of each of the four crystals to their data-capture sheets and describe them. They will need to use magnifying lenses to examine the crystals and complete their report.

Extensions

Have students prepare sample cards of the original chemical form of the salt, alum, Epsom salt, and cupric sulfate solids as follows:

*   Cut a hole in the center of a 3" x 5" (7.5 cm x 12.5 cm) file card.
*   Cover the hole with transparent tape.
*   Press the tape into the powdered solid.
*   Label each card with the name of the chemical.

*   Make enough sets of cards for eight groups of students.
*   Let students compare their "home grown" crystals with those of the original salt, alum, Epsom salt, and cupric sulfate used to make their solution. They will find these to be the same shapes. Explain that the crystal shape depends upon the type of chemical, but any one chemical will always form the same crystal shape.

Closure

In their rocks and minerals journals, have students write a story about finding the largest crystal in the world.

Crystal Shapes

Salt

Cupric Sulfate

Epsom Salt

Alum

Crystal Creations *(cont.)*

Fill in the information needed below.

Salt Crystal Sample

Natural color: _____

Color added: _____

Drawing of salt through a magnifier

Cut out the rectangle below and tape a sample of salt crystals there.

Alum Crystal Sample

Natural color: _____

Color added: _____

Drawing of alum through a magnifier

Cut out the rectangle below and tape a sample of alum crystals there.

Crystal Creations *(cont.)*

Fill in the information needed below.

Epsom Salt Crystal Sample

Natural color: _____

Color added: _____

Drawing of Epsom salt through a magnifier

Cut out the rectangle below and tape a sample of Epsom salt there.

Cupric Sulfate Crystal Sample

Natural color: _____

Color added: _____

Drawing of Cupric Sulfate through a magnifier

Cut out the rectangle below and tape a sample of cupric sulfate there.

Crystal Shapes

Question

What are the shapes of crystals?

Setting the Stage

- Have students look at their drawings on their data-capture sheets from Crystal Creations.
- Ask volunteers to draw the various crystal shapes on the board.
- Tell students that they are going to make a paper model of a crystal shape in this activity.

Materials Needed for Each Individual

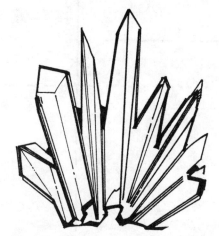

- cubic crystal cut-out (page 54)
- scissors
- crystal samples grown in the "Crystal Creations" activity
- transparent tape

Materials Needed by the Teacher

- transparency of the cubic crystal cut out
- cut-out cubic crystal

Note to the teacher: Students will make a model of the cubic (isometric) shaped crystal and then find an example of this shape among the crystals which they grew in the Crystal Creations activity.

Procedure

1. Use the transparency of the cubic crystal cut-out to show how to fold the shape along the dotted lines and cut along the solid lines.
2. Use the cubic crystal cut-out, demonstrate how to fold it, and tape the edges to form a cube.
3. Distribute the crystal shape cut-out to each student.
4. Have students cut out their crystal shapes along the dark solid lines. Then, have them carefully fold along the dotted lines. Finally, have them use transparent tape to tape the edges of the cube together. Be available so you may help anyone that needs it.

Extensions

Have students examine the crystals they grew and see which one matches the cube they have just made.

Closure

Show students the galena crystals, which are cubic shaped. It may be necessary to break off a piece of the galena to get the cubic shape. Then, in their rocks and minerals journals, have students draw a picture of the cubic shape of a galena crystal.

Crystal Shapes *(cont.)*

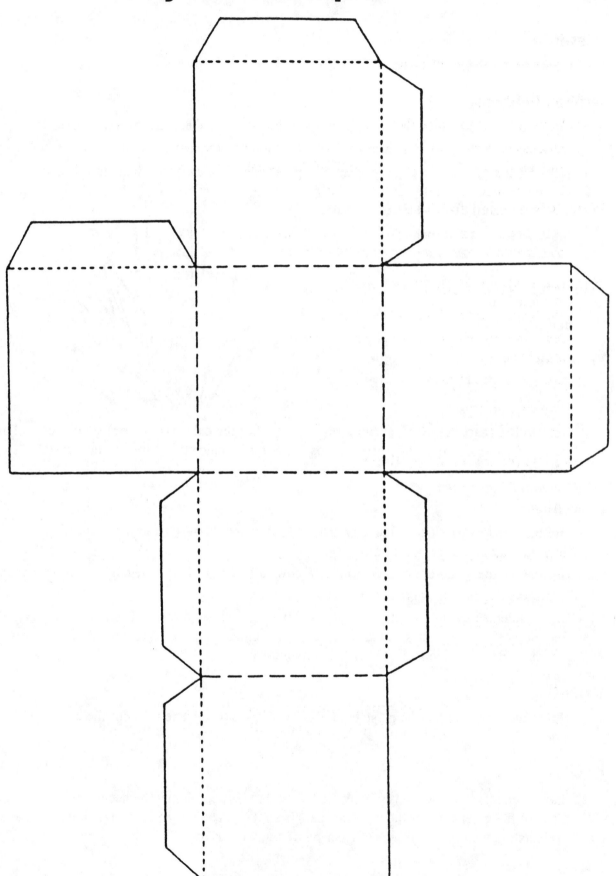

54

Crystal Gardens

Question

What type of crystals form from combining chemicals?

Setting the Stage

- Have students examine the crystals they grew using salt, alum, Epsom salt, and cupric sulfate.
- Discuss with students how they made the solution and then waited until the water had evaporated and left the solids behind in the form of the crystals.
- Remind students that the coloring was added to show how crystals become colored when a different chemical is added.
- Tell students that in this activity they will mix a variety of chemicals to pour over a piece of charcoal or sponge to make a crystal garden.

Materials Needed for Each Individual

- small container such as a margarine tub
- piece of sponge or charcoal about the size of a golf ball
- dropper bottle of food coloring
- magnifying lens
- data-capture sheets (pages 58-59)

Materials Needed by the Teacher

- household ammonia
- water
- salt
- bluing (found in the laundry area of the grocery or drugstore)
- large glass jar for mixing
- measuring cup and spoons
- large spoon for stirring mixture

Teacher Preparation

Note: Do this activity in a well-ventilated area or outside to avoid breathing ammonia fumes.

Mix the following in a large jar to make the chemical solution.

- ¾ cup (188 mL) each of water, bluing, and ammonia
- ½ cup (125 g) salt

Crystal Gardens *(cont.)*

Note to the teacher: A chemical solution will be made using ammonia, salt, water, and bluing. This solution will be poured over charcoal and sponges to create a delicate crystal garden.

Procedure

1. Distribute the charcoal or sponge in the small container to each student.

2. Mix the chemical solution while students observe.

3. Pour 2 tablespoons (30 mL) of the solution over each student's charcoal or sponge.

4. Tell students to add a drop of each different food coloring at a different corner of their sponge or charcoal. Students should leave the middle without any coloring. Then, have students draw a picture of what they observe on their data-capture sheets.

5. Have students leave their containers for 15 minutes in a well-ventilated room or outside in a safe place. Then, have them draw a picture of what they observe on their data-capture sheets.

6. Have students return to where their containers are at the end of the day and draw a picture of what they observe on their data-capture sheets.

7. After 24 hours have students examine their containers and draw a picture of what they observe on their data-capture sheet. Finally, have students use a magnifying lens to draw one of the crystal clusters on their data-capture sheets.

Extensions

Give students a copy of the Crystal Garden Recipe (page 57) for them to try and grow one at home.

Closure

In their rocks and minerals journals, have students write about what they learned when they grew their crystal gardens.

56

Crystal Garden Recipe

You can grow a lovely crystal garden at home. The crystals are delicate and look like white snowflakes. You can add food coloring before the crystals form, and they will be colored red, blue, green, yellow, or a combination of those colors where the drops overlap. This is a good activity to do outside since the ingredients include bluing and ammonia, which have a strong odor.

Materials Needed

- small bowl or plastic cup
- 3 tablespoons (45 mL) each of:
    - * laundry bluing
    - water
    - ammonia
- 2 tablespoons (30 g) of salt (Regular table salt will work well.)
- dropper bottles of food coloring or colored ink
- sponge or charcoal (You may also want to try styrofoam, a tennis ball, or a rough rock.)

  * (can usually be found in grocery stores or drugstores)

Procedure

1. Mix the ingredients together in a bowl or measuring cup and stir to dissolve the salt.
2. Put the dry sponge or charcoal into separate bowls. (This recipe is enough for three or four containers.)
3. Pour the liquid mixture over the sponge, charcoal, or other object.
4. Scatter several drops of food coloring over the solution.
5. Leave the dish in a well-ventilated room or outside.
6. The crystals take about 24 hours to form, less if it is a dry day. They will continue to grow until all the liquid has evaporated. They are delicate crystals which will break if bumped or touched.

Crystal Gardens *(cont.)*

In the boxes below, make drawings of the crystals you see in your crystal garden.

Picture inside container just after pouring solution.

Time: _____

Date:_____

Picture inside container 15 minutes after pouring solution.

Time: _____

Date:_____

Crystal Gardens *(cont.)*

In the boxes below, make drawings of the crystals you see in your crystal garden.

Picture inside container at the end of the day.

Time: _____

Date:_____

Picture inside container after 24 hours of crystallizing.

Time:_____

Date: _____

Simulated Geodes

Question

How are geodes formed?

Setting the Stage

- Show students examples or pictures of a variety of geodes. Explain that geodes are rocks which are hollow inside, often containing crystals which fill this cavity.
- Explain to students that in this activity they will make simulated geodes.
- Review with students the steps used in the Crystal Creations activity for making crystals.
- Explain to students that in order to make geodes, they will use the same solutions used in the Crystal Creations activity but will pour it into coconut shells to represent a hollow rock.

Materials Needed for Each Individual

- half of a coconut shell
- large paper cup of hot water
- salt, alum, or Epsom salt
- red, green, and blue food coloring in dropper bottles
- spoon or stick for stirring
- magnifying lens
- data-capture sheets (pages 61-62)

  Note to the teacher: Simulated geodes will be made from empty coconut shells. These will be compared with geodes or pictures of geodes.

Procedure

1. Distribute materials to the students.
2. Divide students into thirds and distribute the salt, alum and Epsom salt to them.
3. After the minerals have been dissolved in hot water, have students add a drop of food coloring to the solution and then pour it into the coconut shell to fill it.
4. The shells will need to sit for several days until the liquid has evaporated.
5. Have students fill in their data-capture sheet when they have finished making their geodes.

Extensions

Have students create an exhibit for their geodes and rocks (made in the Homemade Rock activity) and then display them for other students and their parents.

Closure

Have students complete their data-capture sheets and then add them to their rocks and minerals journals.

Simulated Geodes *(cont.)*

Fill in the information needed below.

These are the things which I used to make my geode:

_____

_____

This is how I made my geode:

Step 1: _____

Step 2: _____

Step 3: _____

Step 4: _____

It took _____ days before the crystals began to show.

After _____ days, all the water had evaporated.

Simulated Geodes *(cont.)*

Draw pictures in the spaces provided below.

My geode looks like this:

Top View:

Side View:

When I look at the crystals in my geode through a magnifying lens, they look like this:

Making Crystals

Assessment Activity

Question

How can crystals be made?

Setting the Stage

- Ask students to think about the crystal-growing activities which they have just finished.
- Tell students that they will now design an experience to grow crystals.

Materials Needed for Each Individual

- small clear cup
- shallow dish, petri dish, or small lid
- stirring stick
- data-capture sheets (pages 64-65)

Materials To Be Placed in a Central Supply Station for All Students

- salt, alum, and Epsom salt in labeled cups
- spoon for each of the minerals
- hot water in a container with a spout, such as a teapot

  Note: The water temperature should be hot to the touch but not hot enough to burn the skin.

Preparation of Classroom

- Create private working space for each student. This may be done by using cardboard dividers on each student's desk.
- Distribute the materials each student will need as listed above.
- Set up the central supply station with the materials for students to get as they work.

  Note to the teacher: Students will design their own experience to make crystals. This assessment will take several days, in order to let the crystals develop.

Procedure

1. Read the data-capture sheet to students to be sure they understand the instructions.
2. Explain to students that during the assessment they may not talk unless they are speaking to the teacher, so that everyone will be able to concentrate on their own work.
3. Tell students that they may come to the supply center when they reach part D on their assessment form.
4. Monitor the supply center to assist students in getting their materials, especially the hot water.
5. Evaluate students' skills as they work, checking for understanding of the method for making crystals.

Closure

Have students volunteer to read their experiences aloud. Examine their crystals as they develop and compare the variations in crystals to related experiences.

Making Crystals *(cont.)*

Fill in the information needed below.

A. Circle the mineral you will use for your experience:

      salt   **alum**   **Epsom salt**

B. List all the things you will need to use for this experience:

_____   _____

_____   _____

_____   _____

C. Write exactly how you will do your crystal growing experience:

_____

_____

_____

D. Go to the supply center and get the things you will need to do your crystal growing experience.

E. Follow the directions you wrote above and do the experience.

F. Now that you have finished the experience, what do you need to do next?

_____

_____

G. Over the next few days keep a record of your experience below:

Date   **Description of what is happening**

_____   _____

_____   _____

_____   _____

Making Crystals (cont.)

H. After the crystals begin to form, look at them with a magnifying lens and draw them below:

Language Arts

Reading, writing, listening, and speaking experiences blend easily with the teaching and reinforcement of science concepts. Science can be a focal point as you guide your students through poems and stories, stimulating writing assignments, and dramatic oral presentations. If carefully chosen, language arts material can serve as a springboard to a rocks and minerals lesson, the lesson itself, or an entertaining review.

There is a wealth of good literature to help you connect your curriculum. Some excellent choices are suggested in the Bibliography (pages 95-96). Two of these choices are briefly reviewed for you on this page.

Science Concept: If you traveled from the crust of the earth to its core, you would find different layers which gradually change in temperature and material.

How to Dig a Hole to the Other Side of the World by Faith McNulty (Harper Trophy, 1979).

This is a delightful story of a boy digging into the center of the earth. He finds the need for special devices to help him as he digs through the hard crust, underwater, and into the hot interior.

- Have students write a skit which depicts the problems which would be encountered as they dig into the center of the earth.
- Have students design the type of equipment needed to help in this dig.
- Have students write a log of the journey describing what they find along the way.

Language Arts *(cont.)*

Science Concept: A variety of minerals make up the earth's crust.

The Magic School Bus Inside the Earth by Joanna Cole (Scholastic Books, 1987).

Ms. Frizzle takes her class on another unique field trip, this time to the center of the earth. Students learn about soil, rock formation, the rock cycle, and the use of rocks.

- Have students find out the uses of the eight minerals used in this lesson, as well as other minerals.
- Have students locate minerals which are used at school and home.
- Have students make a mineral collection, including labels for each mineral.
- Have students write descriptions of various minerals, including gems, so someone can "guess" what each is.

Examples:
- My mineral is soft and smooth. I can easily scratch it with my fingernail. When it is ground into a powder, it is soft enough to put on your skin.

  What is my mineral? (talc)
- My mineral is a shiny, green jewel. People like to wear this mineral around their necks or on their fingers. It is rare and very expensive.
- What is my mineral? (emerald)

Social Studies

Minerals have played an important role in history, especially those which are rare, such as gems and precious metals. The discovery of such minerals has often led to sudden development and rapid population growth, as well as the abandoned "ghost towns" left after most of the minerals had been mined.

Look for books which tell stories of the gold rush in California and Alaska, the diamond mines in South Africa, and silver in Arizona and Colorado. Let the students know of the hardships suffered by prospectors looking for these riches.

Science Concept: Minerals have value for different reasons.

- Have students make a map showing the location of gold and silver mines in the United States.
- Have students show where precious gemstones are found on a world map.
- Birthstones are gems which are related to the month of the year. People once thought they brought them good luck. Today people often purchase a birthstone for a birthday gift.
- Have students find the history of the use and beliefs related to birthstones.
- Have students use the list of birthstones below to make a poster with the pictures of these gemstones.

BIRTHSTONES

January	*Garnet*
February	*Amethyst*
March	*Aquamarine or Bloodstone*
April	*Diamond*
May	*Emerald*
June	*Pearl, Alexandrite, or Moonstone*
July	*Ruby*
August	*Peridot or Sardonyx*
September	*Sapphire*
October	*Opal or Tourmaline*
November	*Topaz*
December	*Turquoise or Zircon*

Math

The study of rocks and minerals can be used to give students experience in measuring mass. Use balances and gram masses to determine the mass of the eight mineral specimens from the activities. Let the students compare the mass of different samples of the same mineral to discover that the mass depends upon the size of the mineral. They will also find the mass depends upon the density of a mineral when they compare the mass of two different minerals which are approximately the same size. Galena has the greatest density among these mineral specimens, and talc has the least.

Science Concept: The mass of a mineral depends upon its density as well as its size.

- Have students weigh a sample of each mineral shown below and record its mass.
- Have students compare their charts with others, using the same minerals but different samples.
- Were they the same mass?
- If they are a different mass, how could this happen?

Mineral	Mass
Calcite	_____grams
Galena	_____grams
Graphite	_____grams
Hematite	_____grams
Magnetite	_____grams
Obsidian	_____grams
Quartz	_____grams
Talc	_____grams

List the minerals on your chart in order of their mass:

1. _____   5. _____

2. _____   6. _____

3. _____   7. _____

4. _____   8. _____

These mineral samples are all about the same size. Circle the one with the greatest mass. Why does it have more mass even though it is about the same size as the others?

Art

The Navajo Indians used sand for paintings in their ceremonies, especially for healing the sick. They made colored sands by grinding stones from the rocks near where they lived. Medicine men and women made the pictures in the dirt using designs they had memorized from earlier members of the tribe. The person who was to be healed sat on top of the painting, which was destroyed when the ceremony ended. Colored sand paintings were also made in Japan in 600 A.D. and in England and France during the 1700's and 1800's.

Have students make their own sand paintings using the sand samples gathered for this study. They can add color to the sand by adding dry tempera paint. The paint will spread when it becomes wet from the glue, leaving a background of color under the sand.

Science Concept: Minerals have many uses.

Setting the Stage

- Make a sand painting to show the students. You may wish to use other materials such as rice or dried seeds with the sand to add other colors and textures.
- Show the students the sand painting you have made to inspire them as they make their own designs.

Materials Needed for Each Individual

- samples of different types and colors of sand
- rice and dried seeds (optional)
- white glue
- tagboard or light colored cardboard
- newspaper

Note to the Teacher: Sand paintings will be made using a variety of types of sand. They will be compared with pictures or examples of actual sand paintings.

Procedure *(Student Instructions)*

1. Draw a design on the tagboard for the sand painting, and then place newspaper under it.
2. Spread a thin layer of white glue on the sections of the design which will use the same color of sand.
3. Slowly pour the sand over the glue, gently press the sand into the glue, and let it dry. (It may be best to let each section dry for an hour before beginning the next.)
4. Continue spreading glue and sand until the entire painting is complete.

Extensions

Have students find information about the history and types of sand painting to share with the students.

Closure

Have students create displays and show their sand paintings.

Creative Dramatics

Let students pretend they are part of the earth's crust undergoing gradual changes. Through the following game, the students can physically express what they have learned about the rock cycle.

Science Concept: Rocks and minerals are constantly undergoing change.

Setting the Scene

Make large signs of each of the following terms: (number of signs needed)

igneous (4)	*sedimentary (2)*	*metamorphic (2)*
magma (5)	*volcano (1)*	*mountain (1)*
ocean (1)	*river (1)*	*rain (1)*

Place a table in front of the room and place chairs (X) at each end. Place a large piece of brown butcher paper next to the table as shown:

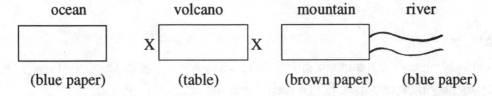

- Tape the ocean, river, and mountain signs to the correct colored butcher paper.
- Cut a door in the side of the mountain near the river for a student to walk through.

Assigning the Roles

- Assign 16 students to roles in the Rock Cycle game as follows:
  * two students hold the ocean at each end.
  * five students are magma.
  * one student holds the volcano sign.
  * two students hold the chairs in place and assist the actors to climb up and down.
  * one student uses a spray bottle of water and holds the sign "rain."
  * five students hold extra signs at the ocean, volcano, and mountain areas as shown:

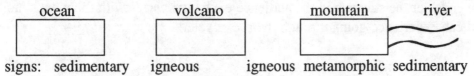

- Give each student playing the role of a rock type or magma a sign to carry.
- Position the two students who are "magma" under the table.
- Place the other three students who are "magma" on the floor under the brown paper.
- Have the one holding the sign "volcano" stand behind the table.
- Let the two students who hold up the ocean and the chair holders take their places.
- Lay the "river" at the foot of the "mountain" as shown in the picture.

Creative Dramatics *(cont.)*

The Action

The Volcano:

- Magma moves from under the table to the chair and climbs on top of the volcano.
- The first "magma" actor crawls across the table to the other end towards the ocean, climbing down off the table and going beneath the ocean. Their "magma" sign is now exchanged for "igneous."
- The second magma actor exchanges their sign for "igneous" and sits on top of the mountain.

The Ocean:

- The ocean is gently raised and lowered to simulate waves.
- The igneous rock continues to move under the ocean throwing small wads of black paper to represent breaking into sedimentary rock.
- The "igneous" sign is exchanged for "sedimentary" and the student sits on the floor under the ocean.

The Mountain:

- The magma slowly raises the mountain up, pushing some areas higher than others.
- The three actors playing magma exchange their signs for "igneous" signs.
- Two actors on either side of an actor who is "magma," exchange their igneous signs for "metamorphic."

The Rain:

- The student sprays a fine mist of "rain" on the side of the mountain nearest the river.
- One student carrying the "metamorphic" sign emerges from the side of the mountain and exchanges the sign for "sedimentary." They continue away from the mountain into the river.

The Audience:

The audience gently claps their hands in a slow rhythm to represent the passage of time.

The Final Play

After the game ends, let the students in the audience exchange places with the players and repeat the game so all can experience going through the rock cycle.

Final Assessment

The activities in this section are designed to assess understanding of the concepts taught in the Rocks and Minerals activities, as well as the Science-Process Skills of:

Observing	Categorizing
Communicating	Relating
Comparing	Inferring
Ordering	Applying

Students will rotate through a series of four activities set up as a center. Multiples of each center are needed to accommodate all the students. This may be done by dividing tables into four areas with cardboard dividers and moving the students around the table as shown below:

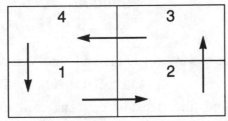

Allow sufficient time at each center for the students to complete the activity and then rotate the students at the same time. Approximately 15 minutes is required for each activity. Those students who finish early should remain at their center and can be provided with a book or science magazine to read while waiting for the others.

Instructions for the Assessment: *(Read the following aloud to the students.)*

- Today, you will do four activities so I can see how much you have learned from our study of rocks and minerals. Each of you will work on the activities at a center and fill in an answer sheet. You will not all begin at Center 1. The number of your center is on the table. You will need to write the answers for the activity at this center on your answer sheet. (Read the answer sheet aloud so the students will know what to do at each center.)

- I will give you about 15 minutes to do each activity. Then I will explain how to move to the next center. It is very important not to talk as you work, so everyone can do their best work. If you have questions as you work, raise your hand and I will come to help you.

- Do you have any questions before we begin? Find the number of your center on the table and circle it on your answer sheet. (Monitor the students to be sure they are on the correct place on the answer sheet.) You may begin working.

Center 1: What are these sand samples?

Preparation of the Center

- Students will be given samples of the four types of sand A - D used in the Teeny Tiny Rocks activity and a mixture of two of these specimens. They will examine the samples of A - D, compare them to the mixtures, and identify the type of sand in each mixture.

Materials Needed for this Center

- magnifying lens
- samples of sand A - D on labeled file cards

  Cut a hole in the card and cover it with clear tape; press the sand on to the tape.

  Sand A

  two file cards with mixtures of sand A - D marked Sand Sample 1 and Sand Sample 2.

- Sand sample 1 should be a mixture of sand specimens B and D
- Sand sample 2 should be a mixture of sand specimens from A and C

  Sand Sample 1

Name_____

You will find these materials at this center:

- one magnifying lens
- four cards marked Sand A, Sand B, Sand C, and Sand D
- two cards marked Sand Sample 1 and Sand Sample 2

1. With the magnifying lens look at the sand samples on the cards marked A, B, C, and D.

2. Look at Sand Sample 1.

   Circle the two sand samples which are mixed together on Sand Sample 1:

   A   B   C   D

3. Look at Sand Sample 2.

   Circle the two sand samples which are mixed together on Sand Sample 2:

   A   B   C   D

74

Center 2: What is the mass of these minerals?

Preparation of the Center

The mass of four of the mineral specimens will be determined by using a balance. The minerals will be listed in order of their mass.

Materials Needed for This Center:

- balance and gram masses
- sample of the minerals: calcite, galena, hematite, and quartz
- Use self-adhesive labels on each mineral to number them 1 - 4.
- Determine the mass of the minerals at each center and record it to check the students' answers following the assessment.

Name_____

You will find these materials at this center:

- balance and gram masses
- four samples of minerals

1. There is a number on each of the minerals. Put them in order from 1 - 4.

2. Find the mass in grams of each mineral and record this on the spaces below.

Mass of the Minerals

Number	Mineral	Mass
1	calcite	_____grams
2	galena	_____grams
3	hematite	_____grams
4	quartz	_____grams

Center 3: What is special about these minerals?

Preparation of the Center

Four minerals will be given to the students. Special properties of only three of the minerals will be described. Students will match the minerals with the descriptions.

Materials Needed for This Center:

- sample of the minerals: graphite, magnetite, obsidian, and talc

    (Mark the minerals A - D using self-adhesive labels.)
- a streak plate and a magnet

- -

Name_____

You will find these materials at this center:

- four mineral specimens marked A - D
- streak plate
- magnet

1. Read the chart below which asks you to find three of the minerals.

2. Find the mineral which matches the description and circle its letter.

Find the Mineral:		Circle the Matching Mineral		
Which mineral can be picked up by a magnet?	**A**	**B**	**C**	**D**
Which mineral could be used as a pencil, since it writes easily on the streak plate?	**A**	**B**	**C**	**D**
Which mineral looks like glass and is very hard?	**A**	**B**	**C**	**D**

Center 4: What crystal is this?

Preparation of the Center

Specimens of the crystals used in Crystal Creations will be used to help students identify crystals grown in a mixture.

Materials Needed for This Center:

- magnifying lens
- samples of the crystals: salt, Epsom salt, cupric sulfate, and alum on labeled cards.
- mixtures of two of these minerals taped to cards labeled Sample A and Sample B:

  Sample A: a mixture of salt and cupric sulfate

  Sample B: a mixture of alum and Epsom salt

Mount specimens of the four crystals and Sample A and Sample B on labeled cards with a hole cut in the center. Place a strip of clear tape over the hole and press the correct sample on to it.

```
┌─────────────────────┐        ┌─────────────────────┐
│        Salt         │        │     Sample A        │
│     ┌────────┐      │        │   ┌────────┐        │
│     │        │      │        │   │        │        │
│     └────────┘      │        │   └────────┘        │
└─────────────────────┘        └─────────────────────┘
```

Name_____

You will find these materials at this center:

- samples of mineral crystals: salt, Epsom salt, cupric sulfate, and alum
- a mixture of two of these mineral crystals marked Sample A and Sample B
- magnifying lens

1. With the magnifying lens, look at the crystals of salt, Epsom salt, cupric sulfate, and alum.

2. Now, look at Sample A and Sample B crystals.

3. Fill in the chart below.

Sample	Circle the two crystals which are in this sample			
A	salt	Epsom salt	cupric sulfate	alum
B	salt	Epsom salt	cupric sulfate	alum

Science Safety

Discuss the necessity for science safety rules. Reinforce the rules on this page or adapt them to meet the needs of your classroom. You may wish to reproduce the rules for each student or post them in the classroom.

1. Begin science activities only after all directions have been given.

2. Never put anything in your mouth unless it is required for the science experience.

3. Always wear safety goggles when participating in any lab experience.

4. Dispose of waste and recyclables in proper containers.

5. Follow classroom rules of behavior while participating in science experiences.

6. Review your basic class safety rules every time you conduct a science experience.

You can still have fun and be safe at the same time!

Rocks and Minerals Journal

Rocks and Minerals Journals are an effective way to integrate science and language arts. Students are to record their observations, thoughts, and questions about past science experiences in a journal to be kept in the science area. The observations may be recorded in sentences or sketches which keep track of changes both in the science item or in the thoughts and discussions of the students.

Rocks and Minerals Journal entries can be completed as a team effort or an individual activity. Be sure to model the making and recording of observations several times when introducing the journals to the science area.

Use the student recordings in the Rocks and Minerals Journals as a focus for class science discussions. You should lead these discussions and guide students with probing questions, but it is usually not necessary for you to give any explanation. Students come to accurate conclusions as a result of classmates' comments and your questioning. Rocks and Minerals Journals can also become part of the students' portfolios and overall assessment program. Journals are valuable assessment tools for parent and student conferences as well.

How To Make a Rocks and Minerals Journal

1. Cut two pieces of 8 ½" x 11" (22 cm x 28 cm) construction paper to create a cover. Reproduce page 80 and glue it to the front cover of the journal. Allow students to draw rocks and minerals pictures in the box on the cover.

2. Insert several Rocks and Minerals Journal pages. (See page 81.)

3. Staple together and cover stapled edge with book tape.

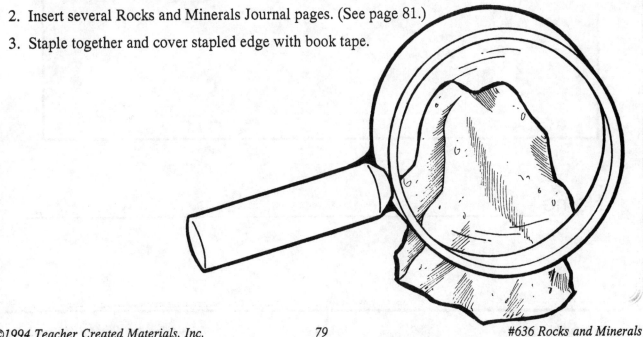

My
Rocks and Minerals
Journal

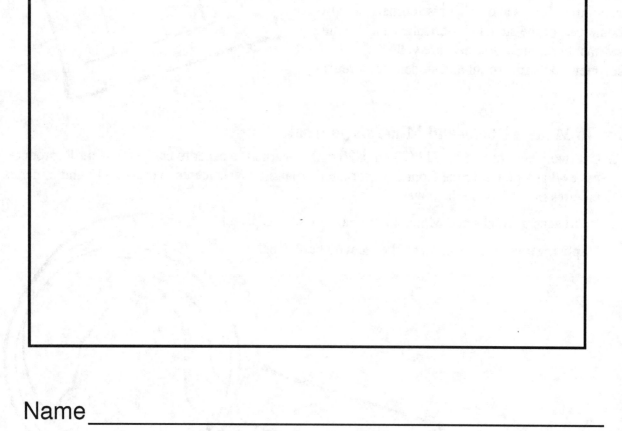

Name_____

Rocks and Minerals Journal

Illustration

This is what happened: _____

_____

_____

_____

_____

This is what I learned: _____

_____

_____

_____

My Science Activity

K-W-L Strategy

Answer each question about the topic you have chosen.

Topic: _____

K - What I Already **Know:** _____

_____

_____

_____

_____

W - What I **Want to Find Out:** _____

_____

_____

_____

_____

L - What I **Learned after Doing the Activity:** _____

_____

_____

_____

_____

Investigation Planner *(Option 1)*

Observation

_____

_____

Question

_____

_____

Hypothesis

_____

_____

Procedure

Materials Needed:

_____

_____

_____

_____

Step-by-Step Directions: (Number each step!)

_____

_____

_____

_____

_____

_____

Investigation Planner *(Option 2)*

Science Experience Form

Scientist _____

Title of Activity _____

Observation: What caused us to ask the question?

_____

_____

_____

Question: What do we want to find out?

_____

_____

_____

Hypothesis: What do we think we will find out?

_____

_____

_____

Procedure: How will we find out? *(List step by step.)*

1. _____

2. _____

3. _____

4. _____

Results: What actually happened?

_____

_____

_____

Conclusions: What did we learn?

_____

_____

Rocks and Minerals Observation Area

In addition to station-to-station activities, students should be given other opportunities for real-life science experiences. For example, volcano models and mineral samples can provide vehicles for discovery learning if students are given time and space to observe them.

Set up a rocks and minerals observation area in your classroom. As children visit this area during open work time, expect to hear stimulating conversations and questions among them. Encourage their curiosity but respect their independence!

Books with facts pertinent to the subject, item, or process being observed should be provided for students who are ready to research more sophisticated information.

Sometimes it is very stimulating to set up a science experience or add something interesting to the Rocks and Minerals Observation Area without a comment from you at all! If the experiment or materials in the observation area should not be disturbed, reinforce with students the need to observe without touching or picking up.

Assessment Forms

The following chart can be used by the teacher to rate cooperative learning groups in a variety of settings.

Science Groups Evaluation Sheet

Room: _____   Date: _____

Activity: _____

Everyone

	Group									
	1	2	3	4	5	6	7	8	9	10
. . . gets started.										
. . . participates.										
. . . knows jobs.										
. . . solves group problems.										
. . . cooperates.										
. . . keeps noise down.										
. . . encourages others.										

Teacher comment

_____

Bragging rights for the group session: _____

_____

Assessment Forms *(cont.)*

The evaluation form below provides student groups with the opportunity to evaluate the group's overall success.

Cooperative Group Evaluation

Assignment: _____

Date: _____

Scientists	**Jobs**
_____	_____
_____	_____
_____	_____
_____	_____

As a group, decide which face you should fill in and complete the remaining sentences.

1. We finished our assignment on time, and we did a good job.

2. We encouraged each other, and we cooperated with each other.

3. We did best at _____

   _____ .

4. Next time we could improve at _____

   _____ .

Assessment Forms *(cont.)*

The following form may be used as part of the assessment process for hands-on science experiences.

Science Anecdotal Record Form

Date: _____

Scientist's Name: _____

Topic: _____

Assessment Situation: _____

Instructional Task: _____

Behavior/Skill Observed: _____

_____

_____

_____

_____

_____

This behavior/skill is important because _____

_____

_____

_____

_____

_____ .

Creating Science Projects

At the end of each lesson in this book, have students think about questions that were left unanswered or that they would enjoy investigating further. Help them focus their questions into science project investigations. The following example shows how the process that is used throughout the book may be used in the creation of science projects.

Example

Teacher: What did you learn in our science lesson today?

Student: How crystals are formed inside of igneous rocks.

Teacher: What other questions about crystal or rock formation would be really interesting to you?

Student: I want to know if the cooling rate in igneous rocks affects the size of crystals that form inside the rocks.

Once students decide which question they would like to investigate, have them use the scientific method to do it. A project stemming from the above question may end up looking something like this:

Question

Does the cooling rate in igneous rocks affect the size of the crystals found inside them?

Hypothesis

Having different cooling rates affects the size of the crystals found inside them.

Materials Needed for Each Group

- hot plate
- small bowl
- measuring spoons
- ice
- ⅓ cup (80 mL) of water
- 3 tbsps (45 g) of cocoa
- large and small pie tin
- cooking pot
- stirring spoon
- butter or vegetable shortening
- 1 cup (225 g) of sugar
- 1 tsp (5 mL) of vanilla

Procedure *(Student Instructions)*

1. Fill the large pie tin with ice.
2. Grease the inside of the small pie tin with butter or shortening and then place firmly in the large pie tin.
3. Mix the cocoa, sugar, vanilla, and water in your cooking pot.
4. Place your cooking pot on the hot plate and boil the mixture for three minutes. Make sure you stir continuously.
5. After your mixture has boiled for three minutes, turn off the hot plate. Pour ⅓ of the contents in the small pie tin to cool over the ice. Pour ⅓ of the contents in the small bowl and let cool on a table. Leave the remaining ⅓ in the pot and let cool on the hot plate.
6. After the mixtures have cooled for at least 30 minutes, examine the mixtures for differences. Draw a picture of each mixture on your data-capture sheet.

Creating Science Projects *(cont.)*

Conclusion *(Circle your conclusion.)*

The results show my hypothesis is *(circle one)*

supported   unsupported.

Having different cooling rates *(circle one)*

does   does not

affect the size of crystals inside them.

Pie Tin:

Small Bowl:

Cooking Pot:

Super Geologist Award

This is to certify that

_____
Name

made a science discovery!

Congratulations!

_____
Teacher

_____
Date

Supply Request Letter

Dear Parents,

Our class will be involved in many science experiences this month. Could you help to make these experiences a success by looking around the house for the items listed below? If so, please send them in with your child as soon as possible.

Thank you!

Glossary

Conclusion—the outcome of an investigation.

Control—a standard measure of comparison in an experiment. The control always stays constant.

Crust—rocky covering around the earth.

Crystal—a solid composed of atoms arranged in an orderly pattern.

Cupric—the chemical form of copper.

Dirt (soil)—mixture of crushed rock and pieces of organic material from plants and animals which covers some areas of the crust.

Element—a substance such as iron, oxygen, and gold which cannot be separated or broken down by ordinary chemical methods.

Experiment—a means of proving or disproving an hypothesis.

Geode—a hollow rock in which crystals have formed inside the cavity.

Hardness—the amount of scratch resistance on a mineral's surface.

Hypothesis(hi-POTH-e-sis)—an educated guess to a question which you are trying to answer.

Igneous Rock—rock formed from melted rock.

Investigation—observation of something followed by a systematic inquiry in order to explain what was originally observed.

Lava—melted magma from beneath the earth's crust which is forced up through cracks and pours forth from volcanoes.

Luster—the amount of light reflected from the surface of a mineral.

Magma—molten rock beneath the earth's crust. The magma layer is about 1,800 miles (2,880 km) thick and about 1600° F (871° C).

Metamorphic Rock—rock which is changed physically and sometimes chemically when subjected to heat and pressure.

Mineral—substances which were never alive, made of one or more elements.

Mineralogist—scientist who studies minerals.

Nebula—cloud of gas and dust-sized pieces of rock and metal.

Observation—careful notice or examination of something.

Plates—large pieces of the earth's crust which fit together like a puzzle.

Procedure—the series of steps that is carried out when doing an experiment.

Question—a formal way of inquiring about a particular topic.

Glossary *(cont.)*

Results—the data collected after performing an experiment.

Rock Cycle—gradual and continuous change of rock in the earth's crust to igneous, sedimentary, or metamorphic rock.

Scientific Method—a creative and systematic process of proving or disproving a given question, following an observation. Observation, question, hypothesis, procedure, results, conclusion, and future investigations.

Scientific-Process Skills—the skills necessary to have in order to be able to think critically. Process skills include: observing, communicating, comparing, ordering, categorizing, relating, inferring, and applying.

Sedimentary Rock—crushed rock and organic material layered and compacted into new rock.

Streak Color—the color a mineral leaves when rubbed across an unglazed white tile.

Texture—the feel of the surface of a mineral.

Variable—the changing factor of an experiment

Bibliography

Arnold, Caroline. *A Walk on the Great Barrier Reef.* Carolrhoda Bks, 1988.

Baines, Rae. *Rocks & Minerals.* Troll Assocs., 1985.

Baylor, Byrd. *Everybody Needs a Rock.* Macmillan, 1974.

Blueford, J.R. *Rock Cycle - Understanding the Earth's Crust.* Math Sci Nucleus, 1992.

Carratello, John and Patty. *Our Changing Earth.* Teacher Created Materials, 1988.

Cole, Joanna. *The Magic School Bus Inside the Earth.* Scholastic Books, 1987.

Fejer, Eva and Fitzsimons, Cecilia. *Rocks and Minerals.* Longmeadow Press, 1988.

Gans, Roma. *Rock Collecting.* Harp J., 1984.

Harshman, Marc & Bonnie Collins. *Rocks in My Pocket.* Dutton Child Bks., 1991.

Hyden, Tom & Anderson, Tim. *Rock Climbing Is for Me.* Lerner Pubns., 1984.

Jennings, Terry. *Rocks.* Marshall Cavendish, 1990.

Jennings, Terry. *Rocks and Soil.* Childrens, 1989.

Kehoe, Michael. *The Rock Quarry Book.* Carolrhoda Bks., 1981.

Macaulay, David. *Underground.* Houghton & Mifflin, 1983.

Marcus, Elizabeth. *Rocks & Minerals.* Troll Assoc., 1983.

Mariner, Tom. *Rocks.* Marshall Cavendish, 1990.

McNulty, Faith. *How to Dig a Hole to the Other Side of the World.* Harper Trophy, 1979.

Mitgutsch, Ali. *From Ore to Spoon.* Carolrhoda Bks., 1981.

Podendorf, Illa. *Rocks & Minerals.* Childrens, 1982.

Richardson, Joy. *Rocks & Soil.* Watts, 1992.

    Rocks & Minerals. Price Stern, 1987.

    Rocks & Minerals. Eye-Witness Books Series. Alfred Knopf, 1988.

Seddon, Tony & Bailey, Jill. *Physical World.* Doubleday, 1987.

Selsam, Millicent E. & Hunt, Joyce. *A First Look at Rocks.* Walker & Co., 1984.

Sipiera, Paul. *I Can Be a Geologist.* Childrens, 1986.

    The World Of Rocks and Minerals, Young Readers Press, Inc., 1971.

Thurber, Walter, et al. *Exploring Earth Science: Geology.* Allyn and Bacon, Inc, 1976.

Zim, Herbert. *Rocks and Minerals.* Golden Press, 1957.

Spanish Titles

Argueta, M. *Los Perros Magicos De Los Volcanes (Magic Dogs Of The Volcanoes).* Children's Book Press, 1990.

Brown, M. *Sopa De Piedras (Stone Soup).* Lectorum, 1991.

Cole, J. *El Autobus Magico En El Interior Da La Tierra (Magic School Bus Inside The Earth).* Scholastic Book Services, 1987.

Munsch, R. *Agu, Agu, Agu (Murmel, Murmel, Murmel).* Firefly Books, 1989.

Steig, W. *Silvestre Y La Piedrecita Magica (Sylvester And The Magic Pebble).* Lectorum, 1990.

Bibliography *(cont.)*

Technology

Agency for Instructional Technology. ***Rocks: When Is a Rock a Liquid?*** and ***Let's Explore Soil and Rocks.*** Available from AIT, (800)457-4509. video

Bill Walker Productions. ***Where Does Sand Come From?*** and ***What Good Are Rocks?*** Available from Cornet/MTI Film & Video, (800)777-8100. videodisc

January Productions. ***Rocks and Minerals.*** Available from CDL Software Shop, (800)637-0047. software

Learning Corporation of America. ***How To Dig A Hole To The Other Side Of The World.*** Available from Cornet/MTI Film & Video, (800)777-8100. video

Lumnivision. ***Gems & Minerals.*** Available from Educational Resources, (800)624-2926. laserdisc

Resources

Carolina Biological Supply Company, 2700 York Road, Burlington, NC. 27215-3398 or P.O. Box 187, Gladstone, OR 97027-0187 (800) 334-5551

Supplier of Biology/Science Materials

Delta Education, Inc., P.O. Box 950, Hudson, NH 03051, (800) 442-5444

Science equipment especially suited for elementary grade levels. Request Elementary Science Study (ESS) catalog for mineral specimens.

National Science Resources Center. Science for Children: Resources for Teachers. National Academy Press, Washington, DC, 1988.

A resource guide for elementary teachers in the area of science, providing descriptions of effective hands-on, inquiry based programs.

National Science Teachers Association (NSTA), 1742 Connecticut Ave., NW, Washington, DC 20009-1171 (202) 328-5800

Membership benefits include subscription to one or more NSTA journals of teaching science which offer many classroom activity ideas. Members also receive NSTA Reports and an extensive annual catalog of science education suppliers.

Ward's Natural Science Establishments, Inc., 5100 West Henrietta Rd., P.O. Box 92912, Rochester, NY 14692-9012 (800) 962-2660

Supplier of materials and equipment for all science areas. Request catalog for geology.